MIXED HARVEST

MIXED HARVEST

Stories from the Human Past

Rob Swigart

berghahn
NEW YORK · OXFORD
www.berghahnbooks.com

First published in 2020 by
Berghahn Books
www.berghahnbooks.com

Library of Congress Cataloging-in-Publication Data

Names: Swigart, Rob, author.
Title: Mixed harvest : stories from the human past / Rob Swigart.
Description: First edition. | New York : Berghahn Books, 2020.
Identifiers: LCCN 2019042445 (print) | LCCN 2019042446 (ebook) | ISBN
  9781789206111 (hardback) | ISBN 9781789206203 (paperback) | ISBN
  9781789206128 (ebook)
Subjects: LCSH: Agriculture--Origin. | Agriculture, Prehistoric. | Human
  settlements. | Land settlement patterns, Prehistoric.
Classification: LCC GN799.A4 S95 2020 (print) | LCC GN799.A4 (ebook) |
  DDC 630.93--dc23
LC record available at https://lccn.loc.gov/2019042445
LC ebook record available at https://lccn.loc.gov/2019042446

British Library Cataloguing in Publication Data

A catalogue record for this book is available from the British Library

ISBN 978-1-78920-611-1 hardback
ISBN 978-1-78920-620-3 paperback
ISBN 978-1-78920-612-8 ebook

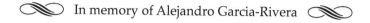 In memory of Alejandro Garcia-Rivera

Humans lived in a realm composed only of other beings; there were no things.

—Bruce Trigger

Harvest follows planting. Among the benefits are soaring cathedrals, orchestral music, printed books, long-form television series, chocolate mousse. Unintended consequences include plagues, pollution, climate change, overpopulation, inequality, and war. Thus, the harvest is mixed.

Some of the people and places in this book are real. Others are fictional. The difference is trivial.

Contents

III. Home

PREFACE

Mixed Harvest is a collection of stories about the deep past and those who lived through millennia of exploration, hardship, and uncertainty during the evolution of farming. In the space of a few thousand years agriculture dominated the earth. We live with it all around us. History began, cities soared, the landscape was crisscrossed with roads.

These stories were inspired by a trip to the Neolithic settlement of Çatalhöyük, which began with a total solar eclipse and ended with a devastating earthquake. Both were natural consequences of celestial physics and plate tectonics, explanations unavailable in the deep past.

A first encounter with Çatalhöyük can horrify: hundreds of small, cramped dwellings were built right next to each other; they are gloomy, darkened by smoke. The only opening was a hole in the roof which served as both entry and chimney. The ribs of skeletons were blackened by carbon. I was aware of my own prejudices in favor of twenty-first century comforts, but I couldn't help wondering why anyone would consider living in such conditions.

A series of seminars at Çatalhöyük on the role of religion in the emergence of the city introduced a more complex and nuanced understanding. This agglomeration of small dwellings on the Konya plain in Anatolia did not appear out of nowhere; it was the consequence of many stories, many decisions that came before. Still more stories are buried here, and many more followed. Collectively they make up an imagined mosaic of our deep past and the myriad of small changes that brought us today's crowded and troubled world.

Religion continues to play a role. So does climate, with its rhythms both natural and unnatural. Intelligence and dexterity at making tools gave the human species a confidence that all too often has led to arrogance. People are adaptable and clever. They migrate and alter, organize, cooperate, and fight. They have remade the world, and in so doing changed everything about themselves, their diet, ways of seeing and understanding the world, how families and societies are organized, what they believe.

The stories here describe how this happened in Europe and the Near East. They do not intend to suggest that human prehistory occurred only there: life was gathering a very different set of experiences in Australia and Asia. Nonetheless, the arc of prehistory still bent, to one degree or another, toward farming.

Archaeology and prehistory are moving targets. New finds, data, and interpretations appear almost daily. Not long ago the disappearance of the Neanderthals was dated to twenty-nine or thirty thousand years ago. Now some push that back to more than forty thousand. Modern Humans first appeared in Europe forty-five thousand years ago, or sixty thousand, or, in the Near East, one hundred and twenty-five thousand years ago, or even one hundred and eighty thousand. Modern humans evolved two hundred thousand years ago or, as more recently suggested, three hundred thousand. More firmly identified dates would not alter the stories. The deep past is, at least for now, exempt from calendric accuracy.

Each story is prefaced by a short introduction and followed by some context in order to stitch the narrative together. Some stories are linked, but most are independent. The stories are gathered into three chapters: "Shelter," "House," and "Home." These represent a progression in where we lived, a series of transformations in technology and consciousness. Once prehistory might have been called a civilizing process, but in light of farming's unintended consequences the term has acquired more complex, less positive connotations.

"Shelter" takes place in Europe and the Near East during the Paleolithic, the Old Stone Age, when anatomically modern humans first encountered Neanderthals, developed large-scale hunting, and painted realistic images of the animals of their world in the depths of nearly inaccessible caves.

"House" moves into the New Stone Age, or Neolithic, in the Near East. People began building square dwellings that lasted for many generations. They planted crops and domesticated animals. The landscape began to change, dotted with the remains of villages, emptier of trees, ever less wild.

"Home" begins with a brief introduction to prepare the way for cities. The stories in this chapter trace the rise of urbanization, organized professional religion, and writing in what is now southern Iraq. Cities grew, and the people who lived in them came to define themselves as city dwellers, members of a special place. Large-scale agriculture managed by complex social structures surrounded the city in ever expanding rings. The wild was driven away for good.

The stories conclude with the emergence of the city. What followed—the rise and fall of empires and the rapid spread of humanity—is well known.

A brief Afterword sums up the meandering course we followed, and what it means today. Above all, *Mixed Harvest* is a plea to pay attention to the past in order to prepare for the future.

ACKNOWLEDGMENTS

Mixed Harvest is dedicated to the memory of Alejandro Garcia-Rivera of the Jesuit School of Theology in Berkeley, California. Alex was an extraordinary example of the human species.

My appreciation goes to those here listed and the many I unintentionally overlooked.

This book exists because Ian Hodder, director of the Çatalhöyük Project, sponsored me as novelist in residence in 2005, and later invited me to participate in a series of seminars on site. Çatalhöyük is an extraordinary, life-changing place; his generosity matched it in both.

Thanks to all the participants in the Templeton seminars, including Victor Buchli, Ian Kuijt, Barbara J. Mills, Peter Pels, Anka Kamerman, LeRon Shults, Wentzel Van Huyssteen, Paul Wason, Mary Weismantel, and Harvey Whitehouse.

Anna Belfer-Cohen and Nigel Goring-Morris of Hebrew University shared many talks, drinks, meals, and laughter both in Turkey and Jerusalem.

Leore Grosman of Hebrew University not only discovered the Snail Creek Shaman, but gave me an unforgettable introduction to Jerusalem's Old City.

Mihriban Özbaşaran, director of the project at Aşıklı Höyük not only invited me for lunch at the Uni but gave me the opportunity to spend time at the home of Lokke and Semmi.

Kimberly C. Patton of Harvard Divinity School was generous with her time and thoughts about what the motherbaby burial meant to her. She put a very human face on the bone.

In both Turkey and Boulder, Stewart Guthrie, author of *Faces in the Clouds*, shared many discussions about religion and the extent to which our ancestors may have spoiled the future, both theirs and ours.

Pre-historian Jean Clottes took me hunting mushrooms in the Pyrenees, shared his long experience with cave paintings, and his profound views on shamanism. He is the exemplar of shamanic flow. Changsar and Farl would not exist without him.

Alan Simmons shared his passion for the pygmy hippos of Cyprus and his hospitality at the fascinating Neolithic hunting camp human beings established there so far from the mainland.

Niek Veldhuis has graciously allowed me to sit in on his Reading Sumerian class at UC Berkeley for many years.

Special thanks to Jeff Kripal, Razor Professor of Religious Studies at Rice University, the most innovative writer about religion I know: inquisitive, inspired, and endlessly imaginative.

Charles Schwalbe Garcia-Lago of Santander introduced me to the caves of Cantabria. Boy and Leaf owe him.

The best reader a book could have is one who engages intensely and constructively, offers substantive suggestions, and whacks typos and dangling participles with equal ferocity. Joey Eschrich of the Center for Science and the Imagination at Arizona State University was that reader.

David Riggs, Emeritus Professor of English at Stanford University, read too many versions with apparent relish and always gave wise and gentle counsel.

Family and friends and writing colleagues inexplicably omitted, my apologies; you shall receive the blessings of the manuscript goddess.

I

SHELTER

Prelude

The Pit of Bones

Our foreparents emerged somewhere along a spectrum of hominid evolution, which includes great apes, chimpanzees, and modern humans. Was it tools that announced our appearance? Cooking? Burial? Speech? We look back at a tapestry of tiny moments: a finger bone, a pair of footprints, small skulls, big toes and upright walking, butchery marks on deer bones, campfire charcoal, and see in them a thousand beginnings.

We will introduce ourselves not with modern humans but with an ancestor species who lived at a real place in Spain called Sima de los Huesos, the Pit of Bones, around three hundred thousand years ago.

Even among his mostly silent people, Sikka was a quiet one. His life had been hard for cycles past remembering, but he endured without noise and seldom made a gesture of complaint.

This day he clung fast to Ser, though she was small, and together they shuffled up the long, gentle slope through dry, frozen grass. There were trees here and there. They were short and bare and bent, huddled against the cold. Some would have bees in the warm season and honey, too. His mouth watered at the faint echo of memory, what would have been.

Before this climb they had crossed the plain where the largest beings roamed, beings with teeth and claws. He and Ser knew them, their shapes and habits, and had made it across by just going ahead, one foot and then the other, not looking about. The beings, the ones with horns, the ones with claws, ignored them.

Sikka and Ser were slow. Well, Sikka was slow. The others were long out of sight.

They stopped to breathe beside a tree, its branches twisted and bare.

Once, he had tumbled from just such a tree. That was on the other side of the river, beyond all the rolling hills. One hand was grasping a branch, the other stabbing down with the long stick he had sharpened and then hardened in a fire. He struck several times at the tiger below, snarling and lashing his tail. The tiger was trying to hurt them, but he had kept it away from the others, above him in the tree. The tiger was bleeding from the mouth and from one eye.

Then he was staring up into leafy branches. So it was summer. The leaves were moving back and forth, back and forth. The others stood around him, looking down.

There was no tiger. He asked with gestures, making sounds in his throat. They lifted their shoulders and let them drop, pointing to the Skyfire, halfway down to a dark wood. The tiger had gone.

The fall had done something to his sight. He squeezed his eyes shut and opened them, but the gray remained. There was no color. Twice already, walking across the plain, he had asked Ser, making his sounds for the things of the world, if Skyfire had gone below. The day was too short.

Each time he asked, she waved. Skyfire was still there. She moved her hand through the breath pluming out of his mouth, sent it flickering away in the cold. The day was as long as it needed to be, no more.

Grass crackled when they took a step. Their bare feet left wet spots like the stones they shaped to cut open the beings they caught. There were five small pools for the toes.

Ser took his chin and turned his head. He could scarcely make out their tracks, faintly dark against the brown and white. The spots must go all the way across the plain, but he couldn't see them and had to believe her when she showed him they disappeared. The wetness turned white and hard again. Her hands moved, shaping the plume, the emptiness, the little pools.

Sikka had a memory of when Ser was born, so small against her mother's breast. Now she was nearly as tall as he. He was smaller now. He was a hunter, a strong male. The others had touched his arms and back when they could. They scratched through his hair. He remembered it felt good. Now he was short and bent. Tired.

That, too, was some yesterday he couldn't name. Today he put one foot in front of the other, leaning on the female whose sound was *Ser*. He made the sound aloud then, and she looked up at him in surprise. "Sikka," she answered, squeezing his hand.

He tried a smile. "Sikka." He sounded like a frog, croaking his own name, a hiss and a click.

She tugged on his arm, urging him toward the woods ahead. They were closer now and he could make out the trees, dark gray against the lighter gray of the sky. Then they were among the rough trunks, and the sounds were more closed, more directionless, muffled, like when an arm is held against the mouth. The smells changed too, at first more earth and dead leaves. Then his nose brought him creature scent, and scat, and wet, cold fur, and the faint tang of fear.

She tugged his arm again. The trees were tall and the air was still, not like the plain where the air was always moving. He was warmer here.

Skyfire vanished behind gray, and the white spots began to fall. He felt them sting his forehead. They fell into the hair around his head and face. His hand came away wet when he touched it.

Now he could also smell the others up ahead. They were close so he straightened his back as best he could and they came to the clearing.

The opening was there. He remembered it now. Some of the others were squatting before it. The rest were inside. He could smell creatures in there too. Bear. They would be sleeping. If he and the others stayed away from the bear, they could all share. The white spots didn't fall inside where the air was still.

Some people lifted their arms to him. He shook his head and squeezed his eyes shut. When he opened them, he saw the bodies.

They were young; men, most of them. He counted one finger for each body; more than one hand, less than two. He saw the wounds too, dark-gray blood against light-gray flesh. They were naked, of course. Others, living, would wear their coverings. If they found a dead being with branches on its head, or managed to kill one, they could get most of a body covering from it. It took a long time to cut the skin away from a carcass. The skin of the head could warm the ears. He wore most of one himself, his head where its had been. Hoofless legs dangled down his chest, so they shared its being.

Ser let him go and he sat near the bodies. He recognized one from last year. The others had carried the dead here. He knew this because they were already soft again. The dead grew stiff and hard, then soft. Then they began to return to the ground. The flesh fell open and sagged, busy with small, dark beings. This season was cold, and slow. They had time. He recognized the slash of a long tooth across the chest and put his finger on the wound. That would be from the big, silent being with long teeth. It helped him to feel the tear through the young one's flesh as if it were his own. The sensation was strangely pleasant; the flesh was still warm. He was going to die soon. No one told him this, he just knew.

Ser brought him a handful of nuts. He pounded open one of the nuts, but when he tried to chew, his teeth hurt and he stopped. Ser saw and put some in her mouth. She chewed for a while and spit into her hand. He took the wad of softened nut gratefully. His teeth didn't hurt this time.

One of the females began a low, continuous sound in her throat. One or two others joined in, and males joined and then all were humming. A female began to clap two stones together, tock, tock, tock-tock. Sikka felt warmer surrounded by others, by the sound, moving up and down in tone, interrupted by brief pauses when everyone took a breath, the pauses becoming more coordinated until all were making complex sounds with stops together, and the clacking stones fell into a rhythm, some people clapping their hands.

This went on for a long time. Sikka felt sleepy and began to doze. From time to time Ser nudged him awake, and he hummed for a while but soon grew drowsy. His chin dropped to his chest. This made his beard puff out so he could see it. It was white, like the spots in the air outside.

It was dark when he opened his eyes. The sound had stopped and the others were sleeping. He heard breathing. Ser was nearby. She made a small sound in her throat. Sikka reached out and instead of Ser he touched the young male beside him. He was dead and did not move. Sikka's head bobbed down three times, and he went back to sleep.

When it began during the night, it was so swift and silent he didn't know it was happening. Only when he woke in the morning did he see the others, their bodies twisted. Pain, he recognized that. He had seen this once before, long ago. Half his people had died in one night. More than half were dead this time. He held up a finger for each dead, as he had done before, and the number of the dead was greater than all his fingers. He thought perhaps the sickness came from the bodies, but he didn't know. They had carried their dead here, drawn to the cave, to the passage below.

He stood, but his head didn't feel right. Shadows moved across the stone walls, huge shadows, hump-backed, four-legged shadows. Yet he felt no fear until he remembered Ser.

He reached down. She was near him. He knew that. He reached down and something happened to his legs. Pain flared in his head, like Skyfire when it rose above the trees, blinding him.

He could hear the others now, shuffling over the dry dirt, the dust. Stones clattered. He stared up into the gray. He tried to turn his eyes. Two unfamiliar males bent over and lifted him.

He wanted to make a sound. He was Sikka, the hunter. He was a silent one, but never this silent.

He was on his back looking up at the sky. It was made of stone, light gray and lumpy. If he could lift his arm, he could touch it.

The stone began to move and he thought the males must be carrying him. He wanted to wave to them, tell them he was not dead, not yet, but no sounds came, no movement. He closed his eyes and let go just before they did.

He felt the rush of air. He was sliding down, head first. The smooth stone hissed against the skin of his back. Then the stone let him go, and he fell into empty air, and then the ground stopped him.

Ser sat by the opening where the others were dropping the dead one by one. They were almost finished, though there were so many. Sikka was gone and she felt something. He had been with her since she could remember. Now he was no more.

After a long time she looked around. The beings still living sat near the opening. She folded down a finger for each one. They fit on one hand.

Three hundred thousand years after these events, workmen blasting a railway cut through a limestone karst in the Atapuerca Mountains near Burgos, Spain, exposed a complex of caverns several "stories" deep. People had lived and died there. Of course, they were not modern humans, but hominins, immediate ancestors called *Homo heidelbergensis*. Like us they lived and died, made and discarded tools, and shared their world with other predators, which for them included tigers and lions.

They disposed of their dead by tipping them into an opening in the floor of the cave over twelve meters deep. The bodies dropped out of sight, though perhaps not out of mind if they believed they were sending their dead back down into the earth from whence they rose. Such a bewilderingly complex set of behaviors around the dead and their disposal was, in fact, a life-saving solution to a serious sanitation problem.

Cats, dogs, elephants, and other animals seek private places to die, but deliberate burial is a fundamental human behavior. We feel the loss of our kind but know how important it is to move them away from the living. We can bury them, offer them to the carrion birds of the sky, secrete them in hidden caches, burn, or expose them.

Such disposal is solemn, spoken of in quiet tones. But spoken it is, in prayer and ritual, in dirge and devotion.

More than burial, language is perhaps the greatest of human behaviors. Its origins are complex and contentious, but Sikka's combination of gesture, sound, and expression, while speculative, is a logical proto-language. The 1989 discovery that Neanderthals possessed a hyoid bone, the U-shaped structure that anchors the tongue, indicates they were capable of speech. Indeed, speech is impossible without it. Even *Homo erectus* had such a bone. Earlier than that, proto-language was everywhere, from the alarm cries of vervet monkeys to the gestures of chimpanzees.

Every day new surprises blossom on our family tree. The more we know, the more questions we have. One thing is certain, though: we would not be here today had it not been for the evolving potentialities of our ancestral primates. The following stories depict episodes in our journey from the Ice Age past to the overheating present. On the way fateful, if unwitting, decisions, among them farming, shaped our path to domination.

Bringer

Two hundred and fifty millennia after Sikka and Ser lived and died, a male and female, both human, both successors to *Homo heidelbergensis*, meet somewhere in northern Europe.

The male was called Traveler because he was the first modern human to reach this place since his ancestors emigrated out of their African homeland, driven by drought, curiosity, or opportunity.

The woman of this odd couple belonged to a cousin species that had shared Europe with Sikka's people for a hundred thousand years or more before *heidelbergensis* disappeared for good.

Her name was Bringer and she was a Neanderthal.

She was best among her people at finding food. When her small band praised and gave thanks she would shake her shaggy head and remind them that she was the same as all Bringers, like her mother and her mother's mother, which was as far back as memory would take her. It was no different with the sun and the moon; they came around again and again.

The day would bring the smells of coming winter, with clouds tumbling down from the north, but when she left the rest of the women with the little ones and climbed up from the camp the night was sharp and clear. Brilliant stars pulsed with life in the heavens.

She made her way along the ridge, pausing from time to time to sniff. Just before dawn the wind rose, gusting through the scattered pines and around outcrops of bare rock, a familiar song. At the end of the ridge she stopped.

There was nothing on the air, no scent of either danger or food, but she was Bringer and would not return until she had found meat for the new mothers. There was always meat to scavenge.

Pale light was seeping through the darkness when it came to her, a brief moment as the chill wind shifted: a strong blood-smell of fresh kill. Deer. It was gone almost as soon as it appeared. She settled, waiting for the wind to play again, and when it did, yes, the killer was certainly a lion.

The trees rustled, and cool pine replaced the slaughter somewhere below her. Now that she knew where it was she had plenty of time. She would allow the lion its right to take the first portion.

Soon, certainly by the new moon, she and her band would have to move away from the ice before the heavy snows, but for now in the mating season there was still food.

The males of her band had gone hunting two days ago. They might find something to kill with their spears, but more often than not it was up to Bringer to find the fresh kills and move in before the other scavengers, the jackals and hyenas, the wolves and great birds, arrived.

She picked her way down the slope in the semidarkness, holding her short wooden shaft with its finely worked flint point ready to stab or slash. If she was careful, though, it wouldn't come to that.

The gusts brought a stronger scent now, and more frequently. She passed through open woodland along the edge of a narrow valley, and could hear a small river tumbling from the mountains to the east where the sun was rising, dimmed by mist. Cloud was spreading, gray and even as the northern ice. The sun was a faint orange disk with little warmth, but it brought her world into view and she murmured a brief chant of gratitude.

Through a break in the willows on the opposite bank of the river she saw the lioness crouched over a fresh kill. The lioness couldn't hear Bringer over the sound of the river, but when the wind shifted the beast looked up, blood dripping from her prominent canines. After a moment she swallowed, a great tongue swiped over her lips, and she lowered herself, chin on her paws, sated, at ease and unafraid. In her eyes Bringer saw only cautious curiosity. The tuft of her tail straightened into the air and swayed back and forth. She seemed to grin.

The stocky woman and the great cat gazed intently at one another.

Then, as if in answer to an unspoken question of her own, the lioness stood, stretched, and, with a languid backward glance, padded away into the scattered pines. For a time Bringer could see the faintly striped body ripple behind the trunks. Then it was gone.

There were boulders upstream where she could cross, but she would be out of sight of her prize, and no matter how brief, she couldn't leave it for a moment. Without hesitation she took off her leather boots, held up her soft leather skirt and waded through the water. She was intent on watching the wood for sign of jackal or hyena and paid no attention to the biting cold flowing around her knees or the sharp stones on the river bottom.

The clearing was silent. If another animal had heard the doe scream or the sound of the kill it was still too far away to matter. Once Bringer began butchering, other scavengers would wait until she had finished, for she was human, and, unlike the lion, they would be wary of her.

She climbed the low bank and turned her attention to the doe. The body was lacerated, one flank half eaten, but enough remained to feed her small band for several days. She was Bringer. She couldn't go to fetch the men, for they were hunting. Bringing the women here, burdened as they were by the little ones, would take all day and such a prize would not remain untouched. She could feel eyes in the trees. By the time she returned with the others there would be nothing left.

She would carry back to her people as much of the meat as she could.

She glanced up at the dim disk of the sun. She could taste ice on the gusts swirling down the slope.

She set to work with her stone blade, cutting the meat from the bone in strips and laying them out on a greasy piece of leather. She worked with her back to the river. Any real danger would approach from the trees on this side. Her pale hands worked deftly, slicing, cutting through the joints, and pulling away strips of sinew. She bundled them all together with the internal organs—the heart, kidneys, and liver—stood up, and slung the heavy burden over her shoulder, leaving her hands free.

Only then, when she turned to cross the river once more, did she see the stranger on the other shore. She stopped, boots in one hand, blade in the other.

The stranger was strange indeed. It appeared to be a fully-grown male, though far too tall, and dark like the night. He held two long, pointed shafts in his right hand.

His surprise was as obvious as hers.

Neither his height nor his sooty color was the strangest thing about him. No, it was his clothing, and the long black braids that trailed down over his shoulders from under a conical leather cap. He wore many layers of leather and overlapping flaps and strips of different kinds of fur mingled with strings of snail shells; he must be awfully hot and uncomfortable under all those wrappings, and the dangling objects would impede him. His coverings made no sense to her. Perhaps it was a female?

Despite all this, though, he seemed quite solid and firm, a being of flesh. He might be a man, after all.

His mouth moved as if he were speaking. She couldn't hear over the sound of the water and shook her head.

Although he didn't seem aggressive, she remained wary.

Only when he placed his two spears on the brown grass and opened his hands did she see how unsteady he was, how his body shook. This puzzled her. He was young and it couldn't be the cold, not with all those coverings.

His hands were wrapped in crude mittens. Using those long wooden shafts must be difficult. In the thirty annual cycles of her life Bringer had known only the members of her band and the handful of others they came across in their wanderings below the ice. Never had she seen someone with such dark skin. And such black hair! Her own was reddish brown, drawn back and chopped off at her neck. Why would he need to cover his head? It wasn't winter yet.

He had a friendly expression, though, what she could see of it. She didn't quite trust his smile, but it was open and puzzled, so she asked, "What are you?"

The stranger frowned, putting a mitten to his ear.

She shook her head. "Of course, you can't hear."

She crossed the river, came close to him, and stared at his face. It was like one of her people, but not the same: his forehead was too flat, his head too narrow, and his nose was long and thin, not wide and compact like hers. White plumes flowed from it with each breath. Surely with a face like that his people would mock him? He had almost no chin!

But his eyes were large and liquid brown. When he looked past her across the river and they widened, she turned to look. A hyena was ripping at the remains of the deer. She shouted, but it paid no attention. A second hyena was trotting down from the trees.

She turned back to him and shrugged.

His answering shrug upset his balance, for he staggered a step or two. He tried to wave it off, but before she could stop him, he sat down hard on the ground. A wave of pain passed over his face.

She bent down and breathed into his nostrils. His eyes rolled up and with a long sigh he toppled back. A spasm passed down his body.

What was she to do? When the day ended, the other women would be waiting. But she hesitated to leave him. Though she owed him nothing, he needed her help, and it was the way of her people to help anyone in need, even strangers. People were few and must be cherished. Even this odd-looking figure was a person.

Strong as she was, she couldn't carry him up the ridge and all the way back to camp. Besides, she had the meat, already freezing in its wrappings, to bring to her people. The men would be out hunting for two or three more days, and the young ones would be hungry.

She straightened him out so when his eyes opened he would see the clouds. She examined his pointed wooden shafts. They were much longer than the ones the hunters of her band used, but the tips were not stone, merely fire-hardened wood. As a precaution she moved them to the riverbank. If it came to a conflict between them, he may be taller but she was certainly the stronger even if he had been well fed and fit.

She had seen skin like this, though not so dark, drawn tightly over cheekbones many times before. He was a man close to starvation.

She collected some fallen wood. She took from her pouch the fire starter tinder she always carried. In moments she had blown it into life and fed it the wood. Aromatic pine smoke shredded away across the river. It disturbed the hyenas and they looked up briefly.

He did not stir when she rolled him onto the bed of ferns she prepared.

She dropped a few rounded river stones into the fire. When they were hot, she used two sticks to move them into the water skin she had slung over a branch, and then she set some strips of meat and a handful of leaves collected from along the riverbank to simmering.

The stranger's breathing was shallow, and so slow she thought for a moment he was about to leave his body. She twisted a lock of the hair growing down to the thick ridge over her eyes, deep in thought.

She dripped warm water onto his mouth. Nothing happened. The sky was now the color and texture of hard mountain stone. The sun was no longer visible. The snows would come earlier than usual, within a few days. They would have to move again as soon as the hunters returned.

He made a sound and opened his lips, licking at the water.

When the stew was ready, she scooped some into her crude wooden spoon and held it under his nose. His eyes flew open with a snort and he stared wildly.

She tilted her head, watching him closely. This tall, shivering being with dark round eyes, scraggly beard, and peculiar coverings was a puzzle. She was no longer certain he was a real being after all. Though he felt solid enough, she knew there were some presences in the world that could take physical form. Such things happened all the time. He could be one of those. It was always best to treat such beings, shape shifters that could appear and disappear, with respect and tend to their needs. One day you might need their help.

He made more sounds with his mouth. There was nothing odd about that, most creatures made sounds. Some animals could speak.

Did these sounds mean he was hungry? She held the spoon to his lips and he lapped at it. Soon he was eating greedily. Yes, he had told her he wanted food.

Finally he pushed the spoon away and fell back with a sigh. Soon he was sleeping again. Only then did she eat.

Toward evening some hard flakes of snow began falling from the gray roof of cloud. She studied the slope of the mountain for caves but could see nothing to shelter them. They would have to spend the night in the open. She gathered more wood and built up the fire. Scattered flakes gathered on the ground. She could hear animals moving in the darkness outside the circle of light.

She was Bringer, yet today she had failed to bring food back to the women. Tomorrow she would bring them deer meat, and along with it this stranger if he did not vanish during the night. Sometimes such creatures would fade away like ice melting. That would give the women a story to tell during the long winter. Perhaps he would stay with them. He was in no shape to travel by himself.

The snow stopped and stillness crept over them. The stranger's eyes were open. He was watching the flickering firelight on her face. A spirit would not have this puzzled expression.

He made sounds again and she realized he was speaking, though unlike others of her kind, his words made no sense. His teeth were small and white against his skin.

As the night deepened, he began to shake. This also was strange, that he would be cold. He was so frail, not like a human at all. She lay down next to him. The top of her head would just touch his chin if they were standing. She moved close and warmed him back to life.

He was strong enough the next day to make the climb to the ridge, and though he slowed her down, she was patient, and by nightfall of the second day they reached the camp. The women and children surrounded him. He was exhausted and endured their poking and prodding and endless discussion well into the night.

The men returned a day later, and a few days after that they set out for the low country. The stranger, whom Bringer called Traveler, stayed with the people for more than a year. He shared their meals, helped with their hunts, crouched with them in rock shelters watching the rain.

Bringer gave birth late the following summer, and for days after that Traveler sat much of the day working at a piece of mammoth ivory with a small stone tool. Whenever she tried to look at what he was doing, though, he hid it from her with a smile, and finally she lost interest and stopped trying.

One day in late spring he was gone. On his sleeping place she found the small ivory object strung on a loop of leather. She couldn't imagine what it was. This irregular lump had no apparent use, no edge for cutting, and anyway, ivory was too soft. She turned it around in her thick fingers, squinting at it, but the curves and rounded protuberances meant nothing to her, though they looked a bit like hanging fruits. Finally she hung it around her neck and soon forgot she was wearing it. After a few seasons the leather gave way and the object fell off.

When dark-skinned, dark-eyed modern humans appeared in Europe around fifty or sixty thousand years ago, they encountered a pale, redheaded, and perhaps blue-eyed people well adapted to a cool, cloudy environment.

Encounters between Neanderthals and modern humans happened regularly as their European descendants have significant Neanderthal genetic markers. They passed down to us their hard-won resistance to pathogens and parasitic worms unknown in the African homeland. We are slowly uncovering their genetic connection to things like cancer and depression. Whatever the ultimate truth, their genes are part of our heritage.

Neanderthals and moderns were physically and culturally adapted to different ecosystems and exploited different landscapes. Small Neanderthal groups hunted with short-range weapons at close quarters in rugged, forested home ranges through which they dispersed slowly. Modern humans spread rapidly through vast tracts of relatively open land like the Eurasian plains, where larger social groups could organize complex hunts using long-range throwing spears.

When Traveler appeared, he was starving and exhausted. He depended, as it were, on the kindness of strangers.

Traveler and Bringer were quite different. He was tall and thin; she, heavy-boned, with short forearms and strong hands. Her light covering of soft leather would be all she would need: her compact body held in the heat and her broad nose kept the cold from clawing into her throat and forehead. The prominent ridge over her brow and a bulge at the back of her head were odd but not unattractive.

Though we can't know for sure, Bringer probably had a language. Although the placement of the Neanderthal larynx may have prevented the full range of modern human speech sounds, Bringer had a hyoid bone and the human variant of the FOXP2 gene, present in many vocal animals like songbirds. Communication between them might have been challenging but not impossible.

Bringer's competence with her environment would have astounded Traveler. He lived in larger groups, while her people worked more intimately with one another and their surroundings. From their tools we know Neanderthals were efficient stone workers and good providers. They were largely but not exclusively meat eaters, while moderns exploited a wider range of available plant resources.

Neanderthal burials hint at their attitudes toward death and treatment of the dead. From them we infer they felt part of a greater world, one that included beings invisible to the eye but present in the effects they produce in the physical world. To Bringer, this stranger may very well have been the manifestation of such a spirit. She would not have found this unusual.

A 175,000-year-old Neanderthal sculpture "garden" deep in a cave in France merely hints at ritual and art. The Neanderthals appear to have been a down-to-earth, practical people, but their nonmaterial arts like music, song, and dance will remain forever invisible.

Traveler, on the other hand, devoted much time carving a figurine like one found in Germany, a small headless woman with large breasts and prominent vulva. Since her head is reduced to a loop for a cord, she was probably worn as an amulet.

The Horse Hunters

After leaving Bringer and her people, Traveler went in search of others of his kind, walking five or six hundred kilometers south before he found them.

Although they hunted reindeer and other large herd animals, the horse was plentiful, and they were very good at catching them.

Tales from Traveler's childhood and vague legends among Bringer's people spoke of great mountains of ice toward the rising sun. Because those forbidding stories haunted him as he walked away from the cold, he stayed well to the west where the land was low and gentle.

The moon, glimpsed through clouds, grew and died two cycles before he stopped seeing his breath in the middle of the day.

The nights grew shorter and his occasional thoughts of Bringer were honey and bittercress at once. Though she had saved him and welcomed him to her bed, he had grown weary of her people's incessant preoccupation with the details of their daily lives, their endless talk of food and mating, their simple tales and songs. He yearned for the stories and songs of his own people. He just didn't know where they were.

A year before, his band had been following the bison to higher country. Along the way they had come across a stag mauled by wolves and had feasted well in the shelter of a rock overhang. A cold rain began at dusk. The wind rose, and white fire broke the night apart so rapidly it seemed like day.

They had crowded together for warmth. The children cried out each time the earth shook and the sky roared. Some time after he fell asleep an evil dream spirit attacked him. He fought but it was too strong and he soared up into a splash of impossible stars, ablaze with cold inner heat. That made no sense, but it was the way of dreams. The spirit shook him violently. He flew over and over through the same darkness, out of which claws and talons reached for him. He was carried as if on a current, barely escaping them, yet a pain that was not pain but an echo or a memory of it filled him. He was looking forward and backward at the same time. Faces stared impassively, many belonging to dead he had known, fathers and uncles, women who nursed him, his many band brothers and sisters. A diminishing tail of such faces faded behind him. At last he spiraled into a pool of darkness and silence.

He awoke alone. They had left him to die.

He would have done the same and not thought about it. He knew they sang the death chant over him before leaving. No doubt they believed he was already melting like tallow back into the great net of life.

His head pounded, his legs shook, and he set out into the damp mist as if following his feet might bring him home. But in his confusion he wandered north. Food was ever more scarce until one day he met Bringer and she fed him life, so he stayed and learned her people's speech.

Then she had given birth. The child was a fine, strong boy with brown skin and black hair that turned reddish in certain light. Traveler believed he had helped make him, though one could never be certain of such things.

But when he told Bringer the story of Snow Woman and Cedar Man, she only stared at him. It was true she was warm and kind, but in the end, she was not of his people and early one morning he left.

The subsequent days had been good to him: he had learned so much from Bringer about finding things he needed from the northern woods he was practically a Bringer himself.

He crossed many rivers, walked amid aromatic pines and across barren steppe. Twice, large herds of red deer fled at his approach. At a lake filled with water birds he trapped a dozen ducks and spent three days smoking their meat and rendering their fat. In this way his journey was good, and his pouch remained nearly full.

Light rain fell on and off, and tatters of cloud drifted around him. Late one afternoon he passed close to the dark opening of a cave near the bottom of a low hill and stopped. No doubt bear and lion wintered inside, for they had left a dry flinty smell, but in summer he was sure to have the cave to himself.

Burned wood on the floor of a broad front chamber told him others had stopped here too, though they were long gone. He expected to find a few discarded flint tools, or at least some work chips, but there was nothing by the fire but the wing bones of a single duck. Whoever was here last had eaten little.

He built his own fire and after eating he spent the early hours of the night scraping and chipping at a lump of limestone. In the morning he finished making a tallow lamp. He filled it with duck fat and, leaving the fire banked against his return, moved into the cave.

Had anyone asked, he could not have said why he did this. He did not question himself. The dark opening called him in, and he went, holding his lamp before him. The irregular walls, the thick, pale hanging spears, the pools of water over white stone, all shimmered in the unsteady light.

He hummed as he moved through chambers and twisting passages, and the sound came back to him in layers, sometimes far, sometimes close, pitched higher or lower, drawn out or clipped and salty. Echoes made a map of his journey in his ears. If the tallow gave out and he lost his sight, he could always hear where he was and find his way back.

Once he came across a lion skeleton in a shallow depression made by a cave bear. There were bear claw marks on the walls and he remembered his dream spirit. He could not tell from the bones how the lion had died or why it was sleeping there in a bear's bed.

Later he heard a different echo filled with the sound of water falling, and he caught a slivery glint through a narrow hole low in the side of the passage, a river tunneling down there through soft stone, a river hidden from the upper world of men. A spirit river.

That night he slept in the entrance again. When sunlight stabbed his eyes, he blinked awake and stepped out into a green and blue dazzle.

A broad valley faded over a ridge and into the rising sun. The clouds were gone, and the sloping hillsides were covered with rock outcrops, loose boulders, and irregular patches of green. The air was absolutely still.

Along the valley floor the summer sedges and grasses had grown tall and lush. Scattered oaks dotted the plain. There were no other trees.

He stood for a long time scanning the dark horizon against the immensity of the sky. Still, he almost missed it. He moved his eyes back, and it leaped at him, a dark line against the blue. Smoke.

Smoke meant fire, and a line like that surely pointed down to a human camp, not burning grass or forest. Besides, there had been no storm recently, no white fire to set things ablaze. His heart rose into his throat and remained there.

He set off at once, taking long strides through the grasses. By noon he had lost sight of the smoke, and climbed the steep slope, the sun sliding down to his right.

From the top of a ridge he spotted the smoke again. He had just started down when he heard the rhinoceros. He melted back into a birch thicket to watch.

She was the largest he had ever seen. She snuffed along the verge of a small creek, tossing her long horn high into the air to chew and lowering her mouth again to crop clumps of sweet grass. He could clearly hear the rumble and huff of her breath. She took a few steps between bites and her waddle along the bank of the creek made him smile. The muscles under her thick pad of reddish fur rippled with each step.

Her calf grazed nearby. It was big enough to have been weaned. Even so it drifted close to its mother and tried to nurse. She knocked him away with a flick of her haunches. She would be ready for mating soon.

The creek flowed down the sloping shelf toward the valley. In the far distance sunlight sparkled off the meandering line of a river.

An afternoon breeze had sprung up, bringing the rhino's heavy wet wool smell. He was safe here. Her eyes were small and her sight dim, the birches offered adequate protection. Only a very angry rhino would try to knock them down just to get at him. She and her calf were unlikely to care about a puny human hidden in the trees.

He watched her move a few steps and dip her head to graze some more.

He could avoid them by backtracking along the top of the ridge, but this would take time and he was eager to reach the campfire before the band moved on. The day was advancing. Better to wait them out.

A warbling high-pitched squeal interrupted these thoughts. The massive female's conical ears turned forward, toward the south. She tossed her head, spewing out a spiral of green-tinted saliva, and broke into a jagged trot across the creek. The juvenile looked up in surprise before lunging after her. They vanished around a rock outcropping and silence fell.

He gathered his food sack and fire-hardened spear and followed. Once out of the shelter of the birches he could see that the breeze had broken the wavering line of smoke toward him. The camp was still some distance away, but he should be able to reach it well before sundown.

He descended onto a plain broken by channels, swells, and tilted outcrops of bare gray rock. As he crossed an isolated slope thick with sedges and scrubby oaks the shadow of the western ridge reached toward him. Above the ridge the line of smoke was dissipating into the hazy air. The camp must be either on top of the hill or on the far side. He started climbing.

At the top a precipice fell away on three sides. Lines of golden light and deep shadow slanted across a series of tilted terraces and tumbled boulders in the valley. The breeze died away and on the far side smoke rose straight up once more. He counted eleven summer shelters. There must be at least thirty people in this band.

The shelters curved in two semicircles around an open space. Two women were tending the fire and a small cluster of children played just outside the west opening. He counted at least twenty women and children. The men must be out hunting.

He shaded his eyes against the declining sun and scanned the lower slopes of the western ridge. They were already in deep shadow and he could see little. He looked toward the river. The men could have gone there for fish, and indeed, he saw a cluster of dark forms against the vivid green. He could imagine their jokes, playful shoves, and slaps on the back if they had been successful.

This could even be the band that had abandoned him, though it was not likely. He did not recognize the women tending the fire.

Even if these were not his people, their camp shared the same east-west arrangement as his, the same shelters, the central fire, and western play area. They knapped their flints just behind the southern curve, facing away from the shelters so that each worker had his own semicircle of stone chips in front of him, just as his people would. He could picture the pale mounds of flint chips growing against the green and gold of the grasses.

It would take too long to go all the way back down the hill to get to the camp, so he began searching for a shorter way. Halfway back along the eastern edge the sides sloped more gently. He was looking for a path down when a boar screeched at him from a thicket and blundered off.

He followed the animal to the edge and found a narrow switchback game trail. A short time later he reached the bottom. It was near dusk and growing cold in the shadow of the hill.

The men had returned, and the people of the band were already seated around the fire when he approached out of the gloom, his hands spread wide.

The man who stood then was an elder, still wide-shouldered, and tall. His grizzled hair, tied with leather to one side, hung over his shoulder, and his smile revealed teeth worn short by time. Though the sounds were strange, Traveler had no difficulty understanding the words, "You might tell us why you wear the fur of the gray wolf, and who are your people."

"I will tell you. I was with Moon Deer People and wore skins like yours," Traveler replied. "But a night spirit took me. When I could see again, my Moon Deer People had left, as was right. I lived and found my way to others two moon cycles north of here. They are short, with strange pale skin and reddish hair like a rhinoceros, and often wear the fur of the gray wolf. They called me Traveler in their language, because they found me when I was wandering alone. It is as good a name as any."

"Very cold, north of here, Traveler."

Now that Traveler's eyes had adjusted to the fire light, he could see faces. None were familiar. "Very cold," he said. "It is better here."

"Better here," the elder agreed. "Many horses by the river, reindeer and bison in winter. You are welcome to our land, Traveler. I am called Savik, speaking this day for Laughing Man Family, for so we call ourselves to remember a great grandfather."

"I am pleased, Savik of Laughing Man Family. I thank you."

Savik continued thoughtfully, "We know of Moon Deer. It is told they wander the Land Beyond the Ridges, toward the setting sun. It is told that Laughing Man and Moon Deer met once in fall, in the days of my fathers. It is told that Marda, second half-daughter of my mother's sister, joined the Moon Deer People."

"I know of Marda," Traveler said. "It is told she joined us and berthed with my third father, Mez. This was before I had a woman."

The elder gestured at one of the older hunters near the fire. "Marda was with Labrin before she joined Moon Deer. She wanted to see the Lands Beyond the Ridges."

Labrin looked up from cleaning a fish and nodded. His white beard was cropped short just below the chin, giving him a lion's face.

"Labrin," Traveler nodded. "It is told a woman called Star left Moon Deer when Marda joined us."

Labrin nodded. "Star was with me until a long-tooth killed her."

"I regret," Traveler said.

"I too regret," Labrin answered. "But such the animal spirits do, to give and to take. It is the way. That was near a cave." He nodded to the north. "I went into the darkness with fire and killed the long-tooth with my spear."

"Perhaps I have seen that lion's bones where a cave bear slept. You enter under a slanted rock. It faced a low hill right of the rising sun."

Labrin nodded. "Yes, that was my lion. I sent him back to his fathers."

Savik made a sign, and they opened a place for Traveler by the fire. When all were settled again, someone handed Traveler a small fish, a welcome change. He ate with relish. When the meal was over Savik announced a hunt for the next day. "Traveler, you will join us?" he asked.

"I will join you."

"That stick you carry may be fine for rabbits and baby pigs, but it will not do much to a running horse. You will need a good blade. Labrin can loan you a shaft and in the morning you can make a blade. You can make a blade?"

"I can. I thank you, Labrin. I will return the favor."

Someone began a song Traveler did not know. The others joined in and he began to catch the ways the melody adapted to Great Bear and Lesser Deer, and how the meaning shifted back and forth between hunting and sex. As they sang, the line of seated people shifted, too, and a woman squeezed in beside him. She had seen more than thirty rounds of seasons and was past her best years, but she had a pleasant smile and after the song she invited, and he agreed, to share her shelter. She told him that Bison was the first word she said, and so he could call her that.

At first light he picked among the lumps of flint for a possible spear point. His breath flew from his mouth and vanished, and he had to stamp his feet to stay warm, yet this was the second day in a row without cloud.

By the time the sun flared its first arc he had selected a solid core that fit well to his hand. He took a strike stone and began to knock several long blades on which to work. Most would crack or break, but a few would be suitable.

One of the other men appeared with a sack of berries and nuts. He offered some to Traveler and they paused to share the meal. His name was Striker.

"Your name says you are a maker of good blades," Traveler said.

The other nodded. They sat cross-legged and faced each other, speaking quietly as they worked. Traveler spoke of his time among Bringer's people, of her camp and how it faced the south where the sun lingered instead of east and west. He described the small figure he had carved from mammoth tusk and left for her.

Striker told him they had killed a mammoth two years before. She was old and tired, so they had managed to kill it in only three days. That was a great victory and the meat had lasted them through most of the winter.

"We haven't seen another mammoth since," Striker said. He thought that for cycle after cycle the mammoth herds had been going to the east. He could not remember a time they had gone any other direction.

"To the mountains of ice, we think," he said. Then there were only stragglers. That old, sick one was the last. Beyond the mountains to the east were mountains of ice or vast seas of water, and places where the trees grew close together. He couldn't imagine such a thing, trees touching one another.

"In the north," Traveler said. "Bringer's people said the trees had been leaving for some time."

"Trees don't like company," Striker agreed.

When he had two good blades Traveler borrowed pitch from Striker and fixed the blades into his borrowed shafts. Striker nodded in satisfaction.

When the sun was high Savik and Labrin fetched them to the eastern gap in the camp. All the men now carried two spears and a flint knife or stone axe tucked under their leather waistbands. Those with long hair had tied it back.

The elders led them in single file along the slope to a natural corral of fallen rock at the base of the hill.

Traveler and one of the younger men helped Savik up a series of shelves to the wall of the cliff. The three of them scanned the plain between the outcrop and the river.

Traveler was first to spot a family grazing in full sunlight. He pointed them out.

"What do you see?" Savik asked. "My eyes, at this distance … ."

"A stallion, four mares, three young," Traveler answered.

The youngster cleared his throat. "That one apart from the others could be a stallion, not a mare. At this angle … ."

"A mare," Traveler answered firmly.

"Good enough." Savik climbed back down painfully. He assigned three of the men to conceal themselves among the boulders.

"OK, go," he said when all were ready.

Traveler and six others approached the horses. They split into two groups a few hundred paces away from them and circled around, guiding each other with elaborate arm gestures. Once they had surrounded the herd they closed in, leaving an open path toward the corral.

The stallion lifted his head and flicked his tail in alarm. He whinnied. Before the others looked up, he spotted one of the men and sprang in motion, headed north. Another man rose from the grass, shouting and waving his arms. The stallion veered south. The hunters, waving their arms, ran toward the horses, driving them toward the escarpment. The animals broke into a gallop in full panic, dazzled by the sun. The men struggled to keep up.

The stallion sensed something and swerved south at the last minute. Labrin, compact and energetic despite his white hair, leaped from a boulder and whirled a spear, yelling terrifically. The stallion veered back into the corral, followed by the rest of the horses.

The hunters concealed in the boulders hurled their spears, the others arrived out of breath, and the slaughter went on for some time.

Once the women arrived, the butchering began. The women crouched beside the bodies and spoke softly to the animals in unison as they cut away their hides and flesh. They spoke of their regret at their deaths, and its necessity. They wished them a good journey back into the world from which they had arisen. There, the women said, their lives would be long; there, they murmured, sweet grass grew tall throughout the year, and there was water and good mating.

Bison, who had taken Traveler into her shelter, glanced at him frequently, grinning proudly.

Clouds swept in again at the end of the day, but the rain held off, and at the feast that night Savik embraced Traveler. "If you so wish," he said, "we will not call you Traveler any longer. From now you travel freely in our company, and we will call you Seer of Horses for your keen eye."

"I so wish," the man who had been Traveler replied.

The cave at Azé, fifteen kilometers north of the famous outcropping at Solutré, is one of the deepest in Burgundy. Inside, along with bear and other cat skeletons, is one of the few cave lion skeletons in France. Wandering through such a cave with only a flickering tallow lamp must have been frightening, although a person with trained hearing might navigate through the darkness by sound alone. Anyone who hums in the darkness in one of the deep caves of France or Spain can clearly detect changes in the quality of the sound.

Neanderthals had preceded moderns in the valley. They came for the herds of reindeer and later horses that flooded seasonally through the valley. They were long gone by the time moderns arrived.

Horses were plentiful at Solutré. Thirty-five thousand years ago men trapped them among the boulders near the cliff. The bones bore patterns of age distribution and butchery marks consistent with hunting, while the presence of scrapers shows people used the hides as well. If they killed too many horses for immediate use, they left the remains to rot. Our species has rarely been a careful steward of the earth, but animals were abundant, seemingly without end, and people were few.

Traveler's horse hunters called themselves Laughing Man Family or Moon Deer People as a way to identify themselves to others. Such names did not yet imply extended family relationships or more complex societies like clans or tribes.

People were still thinly spread and usually on the move, but a band member could easily seek novelty or avoid conflict by joining another band. Since there is little genetic evidence of inbreeding, such exchanges must have been the norm. This was very good for maintaining a robust gene pool and minimizing unnecessary violence.

Strife

In a world alive with quick-tempered rhinos, irritable wild cattle, lions, wolves, and hyenas, there was humanity.

Sometimes conflict flares up. Over our span of time, people have found many ways to deal with it. Evolutionary biologists tell us that males are far more violence prone than females.

But conflict is the foundation of story. In this one, a few thousand years later, a stranger brings it to the Flint River People along the Lot River near the modern city of Cahors some three hundred kilometers southwest of Solutré.

Nyla spent the afternoon chipping into the stone above her head. Her child would separate in little more than a moon cycle and the weight of it pressed on her knees. Still, the stone was soft and easily carved, and in the end she had seven rings from one end of the shelter entrance to the other. Once she had hung up the deerskins to block the worst of the autumn wind, she allowed herself to eat.

At dusk the hides flipped aside. Eelar sat down by the fire with a grunt. Only when he had finished eating did he look at her.

He was so different from Odyu, the father of her first child, the boy who fell from the cliff. Odyu was always smiling and calm. The day Eelar appeared Odyu welcomed him. "Who are your people?" he had asked, and Eelar had muttered that he came from the Lands Below the Ice far beyond the rising sun and that was all he would say about his origins. Odyu did not ask why he had come so far from his people. That was not Odyu's way. He merely said, "New eyes and hands, ears and spear are always welcome among Flint River People."

Eelar had grunted, as was *his* way, and Flint River People made room for him. But speaking with him was never easy, as it was with Odyu. Nyla wasn't the only one who found the stranger's silences peculiar. Perhaps, Odyu had suggested, it was just the way of his people. They had all heard stories of the Lands Below the Ice, but no one really believed them until Eelar joined them. Though it was good, Odyu added, to learn those people were not really giants.

After a month with her people Eelar came to Nyla and wrapped his hands around her throat. He was forceful and determined. A man of few words, yes, but while he held her he invoked the spirits of Air and Water, as was right. He pledged meat and shelter, the strength of his arm, the speed and sharpness of his spear. If there was to be a child, he said, he would pledge these things until it had a name.

This too was right, so Nyla accepted him, though without great joy. Now a child grew within her and instead of being pleased, Eelar glared at her from under his shaggy, dark brows with the two deep grooves between them.

She stole a glance at the string of white shells around his neck. They were like none she had ever seen. At first, she stared openly. They seemed a wonder, the way they glowed among the hairs on his chest like a trail of stars in the night sky, whiter than the cliffs along the river. "Stop looking," he had snarled, raising his hand. "They're not for you."

After that she stole glances only when she thought he wouldn't notice. The magic of those shells, the moon-white glow of them kept pulling her back.

He had the necklace from a woman of Raven Shelter, past the setting sun, so he said. The Raven Shelter woman got the necklace from someone else, who got the shells from someone near the Great Salt, or perhaps beyond even that, because the story dribbled to a stop. The necklace had passed through so many hands. Eelar could not say who had drilled the holes and strung the shells on the leather cord. The shells were part of him now. "I am of the Lands Below the Ice," he snarled. "These shells come from the Great Salt, so they told. I have not seen the Great Salt, but these shells have seen me, and now they *are* me. I am both Below the Ice and Great Salt."

He kept the shells around his neck even when he bathed in the river or when he slept. One time the leather broke and he growled and paced for two days, picking the shells from the floor of the shelter. That was when they were gathering along the river in summer. He found them all and made another string. After that he began to look at her in this dark way, as if she had failed to do it for him, to make the leather cord and string the shells together. As if that were a woman's task.

In spite of her hopes, he ignored her longing looks at the shells. From the Great Salt, she thought, beyond the setting sun, always. The stories said the Great Salt was as far as you could go, and that the Great Salt swallowed the sun every day.

Odyu would never have kept the shells to himself like that. Odyu would have offered them to her. Any one of her people would have offered them. Flint River People gifted to one another. She would have worn them for a time, would have absorbed their moon-bone light, and then would have passed them to someone else. One day the necklace would have come back to her. Such cycles, like the moon cycles, the warm and cold cycles, the birth and death cycles, were the Flint River way.

But such was not Eelar's way. He sprawled there against the sloping rock side of the shelter and picked at his teeth with a long fingernail, touching the shells one by one, over and over. She cleared her throat, but he ignored her, making strange growling sounds low in his throat, as was his habit. She dared to ask him once and he said shortly it was the Song of his people and not her concern. She never asked again.

She started to reach for him, but stopped herself. He often moved as though to strike, though he had actually hit her only three or four times, and then not hard enough to leave a mark. His needs were so different from hers.

Flint River People were sheltered under a long limestone shelf above the river. The shelter was naturally divided into several small chambers or niches, deeper and shallower, like the scalloped edge of a leaf. A narrow ledge was a tall man's height below, and below that the cliff fell down to the river, invisible in the darkness.

She got up. "Where do you go?" Eelar demanded curtly.

She tipped her head at the fire in the next niche. "Talk."

He grunted and returned to his Song.

She squatted by Odyu. "I know he comes from Below the Ice," she said in low tones so Eelar would not hear. "He's still strange, Odyu. He's not like us. He watches. When I go down to the river he follows."

"He fears for the child," Odyu suggested in equally low tones.

"He does not fear for the child," she answered firmly.

"He should not. You are strong, Nyla. What happened to the boy who fell, such things happen in this world. The mothers in the river called him back. You know this. Eelar need not fear for the child you carry."

"It's not that." She nodded toward her fire. "See how he is watching us, as if I should not be talking with you. You, First Father of my first child. As if I have failed to provide. Is it because he is from the Lands Below the Ice that he is so dark?"

Odyu considered this. "It is you who fears him. We cannot allow such a fear to remain among the people. Fear should arise when it must and should fade when the need is gone. He has struck you. I know this, too. I have not spoken of it until now, but this night I will discuss with the others this fear that stays, and this man who provokes it. After we have decided, we will speak with him. He has been with us since the last freezing cycles and now the cold is coming again. His people are far from here. He must listen."

The next day when Odyu and the others came to him, Eelar tucked his chin against his chest, hiding the shells beneath his dark beard, and crossed his arms. "You come from the Lands Below the Ice," Odyu began. "We cannot expect you to know as yet the ways of Flint River. But you appear to us a silly person without gender, one who cannot hold his temper. Such a person is worthy only of laughter. We of Flint River know that males who cannot gentle their anger are weak, and cannot gentle themselves. They let the anger come from the dark spirits and seize them."

"I took the woman," Eelar said. "That is the way of my people."

"You took Nyla because she allowed you to take her; she welcomed you as did we all. Now you may stand in silence while we, Flint River People, laugh. Do not doubt we laugh at you. At whom else should we laugh?"

Firelight danced. Eelar glared around the ring of faces. They were showing their teeth, heads thrown back.

"Silly person," someone said. "Not a man."

Another added, "Weak, silly person, who shames himself."

"Hitter of Women we will call this person!"

Laughter began as a forced gasping, a snickering, amid cries of "Hitter of Women." Jibes and catcalls echoed from the back walls of the shelter, and the laughter grew louder, more continuous, more confident. Guffaws and hoots danced along the length of the shelter, niche to niche. Laughter puffed out the leather hangings over the entrance, allowing cold air to curl inside.

Eelar's eyes were fixed on the floor between his feet. He chewed his lip and his biceps jumped. The laughter came in waves, and with each wave his body flinched. He said nothing but his temples swelled and his face grew darker still.

Odyu pointed at him, rocking back on his heels. "Look at him, this stranger, this Hitter of Women. He cannot subdue the night spirit within himself, so how can he expect the Flint River People to help him? We are not his parent-people. We did not birth him. We did not rear him."

"Hitter of Women," someone called.

Eelar lunged at Odyu, hands outstretched.

The laughter stopped. The women parted to let the men move as one to contain this thing who had just become a stranger. He struggled in their grip, but they held him the way an adult holds a struggling toddler, firmly but without anger.

Nyla watched in horror. She felt the growth shift inside her and wanted it out. This person from the Lands Below the Ice had given her the child, and he was of no worth. He told no stories, only droned his monotonous Song without explanation.

Odyu stood before him. His open, smiling face spoke only of sorrow. "Eelar, you have broken your promise to Nyla, to Flint River People. You threatened one of our band; you attacked me. I do not know about the people of the Lands Below the Ice, your people. Perhaps they allow this, but we do not.

"It is the growing-cold season, and you are alone. You may take with you such meat as you can carry, and your hunting tools, but you will leave this camp today. We will not see you again. We will forget you; we will forget your name; Nyla's child already has other fathers. You will have no word in the naming, no say in the caring. You will not exist. Do you understand? You move on like the reindeer after they have finished feeding, like snow when the warming cycles come. Already you are mist, fading away. If Flint River man or woman sees you, they invoke the right of kill, and surely then the fathers of the forest and the mothers of the river below us will take you back. Be warned, Eelar of the Lands Below the Ice. You no longer share our fire, our hunt, our life. Already I cannot see your face. Already you are gone."

Odyu walked away. The others fell in behind him and Eelar was left alone beside Nyla's fire. He waited but none turned back. He followed, walked among them, and no eyes met his, no tongue shaped words for him. One by one he glared into their eyes. They did not turn away, did not lower their lids. He was a spirit, invisible, unseen, unheard.

The skin around his eyes contracted and hardened. He collected his flint knife, his spears; he stuffed his food pouch full.

Wind howled and rattled the hanging skins at the entrance. He went into the gathering storm. This was not the first time Eelar of the Lands Below the Ice stood alone on a cliff like this and looked down into dark water. It would not, he knew, be the last.

He clenched his hand around the bone handle of his knife. He would be back. That was one thing certain. He would be back.

The first hard flakes of snow fell onto the black surface of the river and disappeared.

An idealized Arcadian paradise during the Paleolithic was neither ideal nor a paradise. It was rustic and sometimes harsh, but life, as always, had both pains and pleasures.

Humans were neither noble savages nor solitary, nasty, short-lived brutes, but clever omnivores adapted to a wide range of environments. They were also gifted natural philosophers and keen observers of their world. They left markings in caves and on walls that showed a deep interest in time, particularly the lunar cycle. They still had to contend with each other, though.

Unfortunately, although the world teemed with danger and resources were unevenly distributed, humans were often profligate with earth's plenty. But were they as profligate with each other's lives? Did they fight and kill over mates or food? Was nature really Tennyson's "red in tooth and claw?" Were humans perpetually at war?

Since the Cambrian explosion, life has been equipped with claws and teeth and armor. Violence of predator and prey are part of our world. In the human world, violence with sharp tool-weapons was a feature of daily life.

Usually the violence was directed at local predators or large walking lockers of animal protein. Was it also directed at other men?

Every member of a band was vital to its survival. Cooperative hunting with other bands was far more beneficial than conflict. Violence would be a foolish way to settle scores.

Since the cost would be high, a prudent band would simply move on. Few convincing examples of mass conflict appear in the archaeological record and they appear to have been local and brief. A consistently high level of violence could easily have ended our stay on earth. Far better to shame and send away a disruptive member.

This doesn't mean violence never occurred.

Judgment

Should someone like Eelar persist in aggressive behavior after sanctions, the band would feel it had to respond more forcefully.
An egalitarian group would not do this lightly.

Nights above the river grew longer. As long as the world outside the shelter either howled or hushed, cold kept the Flint River People close. They had frozen meat and leather sacks of wild grain, nuts, and dried autumn fruits. On clearer days the men and one or two of the women ventured onto the plateau above to hunt reindeer foraging under the snow. Sometimes they were successful. The rest of the time they huddled by the fires. They gave little thought to the stranger from the Lands Below the Ice.

The night Nyla's son separated, a pair of recently named children saw something appear and vanish on the ledge below the shelter. The adults didn't believe them when they came in to tell of it; children so young, even if they did have names and speech, were always seeing things in the night.

But they saw it again the next evening right after Nyla's infant began to cry. "Come," the children called, pulling on Odyu's arm. "It's there again."

He lifted the curtain. "A wolf," he declared. "Come for food, that's all. Just a wolf."

All could see two eyes glittering in the darkness below them just beneath the firelight. The animal flowed along the cliff where the firelight touched the ledge. It moved smooth liquid and silvery like the moon on the river's surface. Its eyes never stopped looking up at the shelter.

Odyu threw down some scraps and in the morning they were gone. That evening when the child cried, the wolf was there again, sitting at the edge of the cliff, sweeping a long red tongue over its lips. Its tail twitched slowly.

"See how pale it is," Odyu murmured. "It's gray, but almost white. A moon wolf."

Nyla sat up to nurse and the child stopped crying. "Moon wolf spirit watches over the new one," she said quietly. "It watches."

Odyu hoped it was such a moon wolf spirit, but what he saw in the wolf's red eyes was no protecting spirit. It was Eelar come back.

Snows lingered late into the spring, and every night the wolf was there. Only when the icicles started dripping and they took down the reindeer hide curtains did the wolf stop coming. "Mating season," Odyu said. "He'll be back in the fall." And that, he thought, is what I fear. But he never spoke of this fear to Nyla.

By the following full moon the world had turned dazzling green. There was a day the women left to forage up one of the many streams tumbling down from the high country. The sun was almost hot, but among the willows it was cooler, and there were frogs, as well as reeds and edible roots. Nyla, baby strapped to her back, went with them.

The males stayed near a temporary camp downstream from the rock shelter. Here the riverbanks were low, and they could sit by the water spearing fish and speaking softly to one another. Sometimes they sang a slow warming-season song of mating animals and green things, a song of plenty, a song to lure the fish from their hidden crannies. When someone landed a fish they all laughed with delight, and by late afternoon many gutted fish were drying on a flat rock.

Day was fading when the women returned, leather sacks laden with early shoots, small tubers, and young frogs and a few turtles to roast over the fire. One of the children pointed upstream across the river at the silvery white shape of the wolf pacing slowly back and forth, staring at them. Its lip was curled.

"Same wolf?" Da'a, one of the women, asked in a hushed voice. "Nyla's moon wolf spirit?"

"Yes." Odyu shuddered. "Nyla! Where is she?"

The women looked at one another. "I thought she went up the stream when the sun was high," one said. "I haven't seen her since."

"No, no," Da'a waved her hand. "She went to that stand of willows where the river widens and gets shallow. I saw her cross on the Fallen Rocks. It's not far. She should be here."

The men grabbed pine torches and set off down to the Fallen Rocks.

They searched in the willows. A frog chorus blended with the uneven perfumes of damp earth, growing things, and the flinty smells of the river. All whispered of night spirits, of the living and the dying all around them. They beat through the willows, calling out for her, but the sounds of frog and water grew louder, drowning out her name. The world did not want her found. The river, the cliffs, the streams and meadows, the willows, they had taken her and would not give her up. Odyu searched in despair.

The others had moved away. Their voices still calling her grew faint. Odyu held his torch up and the shadows jumped, startled. Whenever he turned to look, they fled.

Then a noise froze him. Something stealthy was prowling through the underbrush to his right. He pivoted toward the sound. Something was hunting there, the wolf, perhaps. But why was it alone?

The sound stopped. He waited, but it did not return and the sense of immediate danger passed. He called her name again, "Nyla!"

No human sound replied, only the insistent whispers of the night.

Then, shortly after the moon rose, he heard the child cry.

He found Nyla in a small clearing away from the water. A cut in the center of a dark bruise on her cheek was bleeding, and the back of her head was soft. She was unconscious and hot to the touch. The child was nearby, screaming in pain, fear, and hunger.

One by one the other men arrived.

Odyu stuck his torch in a patch of soft ground and crouched beside Nyla, cradling her head in his arms. He washed away the worst of her blood with water from the river. "We have to take them back," he murmured.

Leaving the unconscious woman to the others he picked up the screaming child. It squirmed in his tight embrace, thrashing tiny fists against soft leather of his tunic. Odyu patted its tense back and crooned over and over, "All good, all good, hush. Odyu is here, you're safe." Eventually the screaming slowed to a steady whimper.

They returned over the rocks and up the river to the camp in a somber line. The others took turns carrying Nyla. Once back in the camp they laid her on a bed of ferns.

Odyu arrived with the baby, still crooning to it about Reindeer Man, about Willow Mother, about Young Aunt New Moon and Old Aunt Full Moon.

A younger boy's still-nursing mother took the baby away to feed. Her own child pestered her, tugging at her sleeve, trying to push the strange infant away. She slapped the boy with her free hand. "Hush. It needs, more than you, little one. You have a name; be a person, not an unnamed thing."

Morning brought a chill mist and little light. The women heated stones in the fire and dropped them into skins filled to bursting with water, frogs and tiny tubers. When the stew was ready they let the smell of it waft past Nyla's nostrils. Her head jerked but she remained sluggish and unresponsive.

The mist began to rise late in the morning, and occasional rays of sunlight shot through the camp only to disappear when more mist, colder and damper now, returned. Beads formed on Nyla's face, merged, and flowed down her cheeks. She moaned.

Her eyes opened and she started up with a cry. Odyu put his hand against her chest and gently pressed her down. She looked around wildly. "Where?" she gasped. She clenched and unclenched her fists.

"Hush," Odyu said. "You're safe."

"I'll never be. The boy?"

"Da'a has him. He's hurt, but not badly. Quiet now."

"No, I … ." Her eyes closed and she fell asleep.

Some time later she was looking at him.

"Eelar," he said.

She nodded. "He … ." Her voice was hoarse.

"Breathe, Nyla," Odyu said softly. "Be easy. He's gone, like the wolf."

"No, he … he was so angry."

"Of course, Nyla. We sent him away, and he was angry. I sensed he was going to come back."

"How?"

"The wolf," Odyu said.

"No, the moon wolf, it was good."

"No, Nyla, that was no moon wolf, it was Eelar. I knew when it came the first time. It was watching with Eelar's eyes. He was waiting for you, to find you alone. He did this."

"Yes." Her voice was like the night sounds, a whisper barely heard. "Suddenly he was there. 'What are you doing here?' I asked. The shells glowed against his chest, but his face was scratched, and he was leaner. He was a person whose companions shun him."

Odyu touched her hair, moving it away from her face. "He abandoned us, Nyla. When he hit you, when he shouted, that was when he left us, when he shunned us. He did not follow our ways."

She tried to lift her hand but could not. "I know." She shook her head. Blood oozed from her cheek. "He was angry, Odyu. It was not fitting. He pointed at the boy and shouted, 'I did not make that!' But we did, he and I. I told him that. I told him, 'You are gone from us and have nothing to say.' I didn't know where Da'a and the others were, so I backed away. I tried to run. I was awkward because of the child. It was crying, so afraid. It wailed louder when he tore it away and threw it. It stopped for a moment and started to cry again. Eelar shouted, 'It was Odyu who made that child.' Then he *was* wolf, Odyu. He struck me, hard, and knocked me down. I was lying on the stones. He just walked away into the willows. I couldn't fight him, Odyu. The men from the Lands Beneath the Ice are strong. I'm so … ." Her eyes closed. She gripped his hand, "You, Odyu, First Father of my first … . So tired. Hold me."

He sat with her through the afternoon and evening and into the night. In the morning the rains came.

He tried to feed her, but she pushed his hand away, whispering words he could not understand. The back of her head was still soft. He cradled it and refused food himself.

For two days the rains fell. She pushed his hand away as she settled into herself.

"She leaves the Flint River People," he told the others.

They lowered their heads and watched with him.

When she had stopped breathing, he placed buckskin over her face. The others were somber as they went about their daily tasks and left the two of them, Nyla and Odyu, alone. It was time to move on to collect some of the remaining summer plants and mushrooms and they had to prepare.

There was Nyla, the empty body of her, the meat and bone. They could just leave her. Next year when they came back to this place by the river her body would be gone, or mostly gone.

But there was the child, the hardest task.

Da'a held it in her lap. She was looking up at those gathered around the fire. Their shelters were down, their scant belongings packed for easy carrying.

"What shall we do?" she asked. "It would be a boy."

"We cannot take it," Odyu said. "You know this. Boy or girl, there is no one to carry it, no one to nurse it. Those among us who can nurse must carry their own."

"What shall we do?" she repeated as if she had not heard. She was looking down into the tiny wrinkled face.

"Come, Da'a." Odyu held out his hand. Reluctantly she took it and he helped her stand. The infant stirred but did not wake.

"We can leave them together, side by side," he said.

"What good will that do?" she demanded, but when he took the child she let it go.

They found a fissure in the rocks large enough to hold them. When they left the child still had not awakened.

Da'a grieved over her abandoned Flint River sister Nyla. She grieved as well over the infant, though it was the spawn of Hitter of Women, an unfinished thing and forgotten.

When summer turned to fall and the birches were shedding red and golden leaves Da'a was too ill to continue walking. They were moving on to a river where salmon came in their multitudes and the Flint River People wanted to be there. But for Da'a they stopped at the base of a range of hills. Here the oldest of the men made a strong birch bark tea, and while it steeped in a leather sack he leaned his head back and hissed at the sky. The sagging flesh beneath his jaws quivered.

Da'a and the others watched in fascinated horror as a plant spirit entered the old man and he became stalk and leaf, stem and root, trembling in a faint breeze. A series of deep sighs breathed out of him. The sighs turned to moans, the moans to screams. The hairs on the back of Da'a's neck rippled.

She knew if she did not get better the others would have to leave her, and so she settled herself to leave for the other world. To her surprise they waited with her, and after three days she was strong enough to move again, and so she went with them.

They approached the river across a broad meadow. Faint mist in the air was rich with the smell of fallen leaves and dying grasses and the soft, wet tang of mushrooms under the scattered shrubs.

Now, it had happened more than once in the Peoples' memory that others found their way to this favored spot, so they weren't surprised to see smoke almost lost in the distance down the river.

If this were a large band, fewer salmon might reach them. But if there were no more of them than the Flint River People, then as long as the salmon came in the same numbers as last year there would be plenty for both.

While the others set up camp, Odyu went downstream to greet them, whoever they were. Best to approach them, for they were already here.

If the malicious spirit of grief had not taken Da'a and held them back, the Flint River People would have been first. Already they would be feasting. But they could not change this; it was the way of life in the world.

He walked along the river's edge. The water ran swiftly to his left, passing him by. A salmon leaped in the center of the water, and then another. Though autumn haze obscured the sun, the air was crisp and cool and tinged with the smells and sounds of living: the run of water over stone, the huff and stamp, scurry and screech of animals, the pungency of decay, the eye-sting of smoke in the air.

He heard the voices, one calming, one raised in anger. The words were strangely shaped, and at first he could not understand their meaning, but he continued walking toward the sounds.

Just around a slow bend in the river he almost walked straight onto a broad shelf of bare ground, a place clearly flooded in the spring. Grass and sedges grew in clumps. Only eight shelters were pitched in a semicircle facing the river; this group was smaller than the Flint River People.

Two males faced each other at the edge of the water. The one with his back to Odyu held a stabbing spear in the air. They might have been discussing the hunt, but something in the way they stood made Odyu doubt this.

The other, a short, stocky man, stood calmly. Odyu could see a broad, smooth forehead gleaming under a leather cap, a black beard, woven into five braids, hiding the chin, and several locks of hair streaked with gray dangling beside his cheeks. The man's bare teeth flashed in the watery sunlight. He was not smiling.

Odyu stepped into the open, holding his empty hands away from his sides.

Several men, women, and children stood in small clusters between the shelters and the river. When Odyu appeared they briefly acknowledged his presence and turned back to the pair by the river. Their leather tunics were sewn in zigzag patterns with quills of the porcupine, an animal common in the south. People from the south were almost always friendly, especially when they wandered this far north. Odyu understood they wanted him to wait, so he spread his hands and smiled.

"It is decided," the bearded man said softly.

The other pointed his spear. "I call on Great Elk," he shouted. "I call on the Long-Tooth of the mountains, on the lives of all who came before me in the Lands Below the Ice to grant my right to stay!"

Odyu knew that voice. Eelar had changed his group but not himself.

The shorter one said, "You have offended all those upon whom you call, save perhaps your people. Great Elk and Long-Tooth do not grant your right. You will leave us."

"I, Eelar of the Lands Below the Ice, will take what is mine!" With a quick thrust of his spear the point tore through his opponent's tunic. Blood flowed.

The others let out a shout of dismay and in a rush knocked Eelar to the ground. He struggled furiously but there were too many.

Odyu stepped forward. "I am a stranger to you," he said quietly, and waited for the angry words to fade away. Eelar hung limp in the arms of two of the southern men.

The others looked at him expectantly. "I'm a stranger to you, called Odyu," he repeated. "I will tell you, I am telling you, you will hear that I know this being, who traveled with my people for many cycles."

One of the men holding Eelar asked, "Why did he leave your people?"

"We sent him away, as you are about to do, but I warn you, I tell you truly that he returned to us and killed one of us, the mother of his child."

"I did not kill her!" Eelar shouted. One of the men kicked him.

Odyu ignored the interruption. "He forced himself on her, he injured her, and she has joined the mothers. This thing that calls itself Eelar is less than a man; he is a beast when it has eaten of those blue flowers in the high meadows," Odyu said. "He has the foam of madness on his lips. We can no longer call him human. He is a danger to all he touches."

The men holding Eelar bound his feet to his hands with lengths of braided sinew and threw him to the ground.

One asked Odyu where his people were.

"Up the river. Listen, I speak truly. We have come here for many cycles of the seasons. You can see there are salmon here already. They will come for many days in their multitudes. There will be more than you can use, and more than we can use. They will be beyond counting."

"Then our groups should join," the man said. "I speak. I am called Atch, and we are called Sand Hunters, for in the south there is an area like this, an area of sand, but larger, wide as the sky, filled with orange and pink birds that wade in the shallows of Salt and it is there we have made our home and hunted birds and fish. But not long before now something happened to the lands, to the birds and fish, and we have been coming north for some cycles looking for new lands, new waters."

Odyu said, "I greet you, Atch of the Sand Hunters."

The other nodded, "If it is as you say, Odyu, as you tell, and there truly will be more salmon coming, and look, already their silver backs break the water, then our groups should join together for a time."

A man called out, "For feasting!" And a woman responded, "And mating." This made all the others laugh, even Odyu.

Atch pointed at Eelar. "And also for judging, for if I understand, both our bands have been harmed by him, and sending him away is not enough."

"This is so," Odyu said. "He has harmed my people; he has harmed yours. We think he was sent away from the People Below the Ice, for he had come far."

"He has been violent," Atch said. "He will this day go farther yet if our peoples so decide."

Odyu agreed, and later that day he returned with the Flint River People and the two groups joined together.

All the time the women of the combined groups were out foraging and the men were waiting by the river to spear the salmon arriving in greater and greater numbers, all the time they shared stories, Eelar glared around him with red moon wolf eyes.

That night they had a mighty fire, and together the two groups discussed Eelar and what he had done. The Sand Hunters were going to send him away as the Flint River People had done, but since he had killed a Flint River woman, many felt letting him live would be unwise.

Da'a jumped to her feet, raised her arms, and growled deep in her throat. All talk stopped. She took a shambling step toward Eelar, cocked her head, thrust it forward, and roared. She took another step and her hands became claws, raking at him so he flinched. She paced in a circle, hopping from foot to foot, and all knew she had become a bear sow feeling the rage of the bear defending her cubs, and that Eelar was the threat.

"Bear is in her," Odyu murmured. "Bear anger is in her."

Da'a danced through the night, and one by one the others joined her until all were moving in a shuffling circle around the fire. Every time Da'a roared and scraped her claws at the air, the others fell back with a despairing moan. Someone was hitting a log with a stick, and the shuffling matched the beat.

The fire died down. Light was seeping into the world like blood from a slow wound when at last Da'a collapsed and the women gathered around her looked up to Atch.

"This is the sign," he said.

Odyu said softly, "Eelar must die."

"It is not easy to kill," Atch replied solemnly.

The women all spoke at once. Eelar's actions called for it, Da'a had danced bear, it was right. Everyone agreed: all who were able would take part in the judgment. Anyone, man or woman or named child old enough to thrust a spear would do so, even after Eelar's life had left his body. They would give his body to the fathers and mothers of the river, and the fathers and mothers would carry the body away, along with his shell necklace and whatever powers it might retain. Only by doing this could they be sure that no matter what form Eelar might take next, he could never return.

Larger fauna like mammoth and rhinoceros were on their way to extinction. Bands came into contact with one another more often. Over time humans were reshaping the environment in small ways: carving hooks into a cave ceiling to hold curtains or building rough stone walls to keep out wind and predators. Burning grasslands to encourage more useful plants and animals. Allowing canines to share their camps.

Wolves came in search of scraps. The more docile by nature began to self-domesticate by remaining nearby. Eventually they were helping humans hunt, and so the dog became an intimate of human life.

Without knowing it, people were beginning to domesticate themselves. In areas particularly rich in resources, like seacoasts or river deltas, they could remain all year, clearing the land around fruit trees, casually weeding stands of grain, watering tubers, and performing other small, evanescent horticultural tasks. Over the millennia to come, decreased mobility and increased human density could only increase confrontations over resources, including sexual partners.

The human brain is a powerful tool for making other tools with which to shape the environment, but its decisions are not always beneficial. Individuals could transgress social norms polished over millennia. Jealousy, irrational rage, fear, a desire for revenge for real or imagined slights, or a need to dominate could provoke confrontation. In such cases, survival depended on wise decisions.

Social sanctions for aggressive or selfish behavior existed among primates long before humans appeared. Animal and human groups can tolerate some free riders or bad actors, but when too many threaten the group's survival they must be stopped. A swift response could avert more disruptive conflict.

Drummer

Over and over the ice crawled down from the north to cover much of central Europe and then retreat. Each time the ice came, the Neanderthals retreated to an ever-shrinking refuge on the Iberian Peninsula a few hundred kilometers southwest of Nyla and Odyu's river. There they struggled to survive.

Eventually there was a last. There is always a last. Like individuals, species too die.

A short, stocky male was hunched over a tortoise shell cradled between his crossed legs. The pair of yearling horse tibia bones in his hands whirred against the domed carapace. Their clacking circled around a central double beat. His eyes were squeezed shut against the sweat pouring from his hairline. The droplets plowed damp streaks over his corrugated brow and through the painted stripes on his forehead and cheeks. His reddish beard was stained with charcoal and white clay. He was part human, part animal; part darkness, part spirit.

His posture favored a twisted right leg projecting at an unsettling angle from his hip. Everyone present knew of his painful hopping walk and the grimace that crossed his broad face when he took a step. Everyone knew his differences, his wrinkled brows and short arms. They knew as well that he had become one of them, and they cared for him as a beloved of their own.

His large upper teeth gripped his lower lip, which writhed in frenzied concentration.

On the opposite side of the low-burning fire stood a taller man with roughly cut dark hair breathing into a crane's delicate leg bone, his long fingers dancing over a series of holes drilled into it. Shrill wailing rose, shivered, and stepped downward only to rise again.

The turtle drum articulated the tock-tock-tock of cranes into flight. The stocky drummer playing it growled. Voice and whistle trilled a crane's mating. Other times the delicate bone shrieked the red deer's screeching cry or the peep-peep-peep of a crane resting, and the drummer would answer with a great bison's growling threat.

Animals passed through the two men conversing with one another. The one with the flute stared into the darkness over the head of the drummer, for it was there in the darkness that the spirits prowled.

An old woman squatting nearby scraped a willow wand against a series of grooves carved on the side of a horse's scapula, forcing the double beat. Her gray hair was greased into a fan behind her head.

Her mate crouched before her, head bent toward the embers. Fire tinted his thinning hair with irregular pulses of red and shadow, pulses of heat, of fear, of hope. He was Seer, the conduit, the opening. This gathering was for him.

In the rare moments of silence between beats, when there was no whistle from the flute and the scraping on the scapula stopped, Seer barked short, harsh rasps so low the woman and the others leaned down to hear. A soft breeze set the leaves in the trees around them shifting and sighing, another pained breath in the night.

The others would look at the two musicians and the trilling clatter, the warble, the scrape-scrape, would start up again. An invisible crowd was gathering out of the shadows, the fire, the breeze, the rasping breath, the trees. They came from the water, from the sky.

Slowly the embers sank back into themselves and pulled over themselves a covering of white dust. A pale blue light began to wash away the stars. The moment the sun appeared the man with the flute flowed to his feet, tilted his head, trilled out a series of shrieks, hopped sideways once, and with one long scream from the flute froze in place, a foot in the air.

He was Crane.

The drummer, still bent low over the tortoise shell, threw aside his drumsticks, scooped some ash from the fire's edge and rubbed it over his face, further blurring the painted stripes.

The quiet stretched, taut as a deer hide tied out to dry under the rising sun. This time, instead of the breeze and the rasping breath, there was only the low gurgle of tide flowing into the estuary.

A circle of light appeared beneath the hills over the water where darkness lingered. It pulsed the same way the embers had changed color on Seer's hair. Like all things, it was alive as the embers had been alive. All could feel this brilliance beating with the man's pulse. This was the way blood throbbed from a deep cut in a deer's neck, the way it throbbed in the head when danger suddenly threatened.

The globe, white mist, floated upward, stretching down into the form of a human being, neither man nor woman. It shifted, became a lion, a bear, or a human with a bison's head. The shining creature stared from the darkness, from the pulsing of its own light.

Everyone there froze in place, at once terrified and elated.

The sun floated higher and the creature faded from sight. People stirred, shook their heads, looked at one another in amazement. Someone laughed nervously, and life came back.

The sun was full in the sky. The day would be a fine one.

Seer arched his back. Cords stood out on his neck and his hands curled into fists. He slumped with a barking cough and drew a long, shuddering breath.

After a moment collecting himself, he looked sideways at the old woman. She set aside her scraper and touched his face with the backs of her fingers. She leaned down and sniffed at his breath.

He pushed her away. "I saw," he whispered. "Winter will be hard, harder than last. Dark spirits there, beyond the sound. We must go south, all the way to the other sea. There we will find some safety."

The old woman sat back with a grunt. She looked up at Crane standing on one foot, flute in hand. "We are grateful to you, Crane." She nodded at the twisted one, "And to you, Drummer." Her voice was soft.

Seer breathed easier now. "I did not see all of us there." He spoke as if to himself. "One was missing."

Drummer lifted the tortoise shell from the hollow of his legs and after setting it aside with a long sigh, slumped forward, eyes closed. After a moment he gave a feeble grin.

Crane put his foot down and shook his head. "Was I ...?"

Drummer held himself up with one short, thick arm. "Spirit was here. Bison and Deer were here."

Crane bowed his head. "Good." He slid his flute into a pouch at his side. "We brought them, the spirits," he said to the old woman and her mate. "Drummer and I, we brought them. It is to them you should be thankful, not to Drummer, not to Crane. Bison brought the knowing. Bighorn Deer brought the knowing. They heard us and they came."

The eleven People present murmured, "They heard and they came."

Crane and Drummer touched their lips with their fingertips and spat onto the cooling fire.

The others, the invisible ones, the voiceless ones, the ones with neither taste nor touch, withdrew from the camp. Only the People remained seated around the cooling ashes.

Seer sipped water from the old woman's wooden cup and gently pushed it aside. "I'm fine," he said. "The darkness is gone."

The old women put the cup on a nearby flat stone and stood. "I greet Deer. I greet Bison," she said. "I thank Crane who flew down from the north to help us. I thank them all, and the others whose presence we felt, who came this night to help."

"Not all of us are there next warm season."

"We do not ask who will not be there," the old woman said gently. "We do not ask. That is for the next cycle." *It could be me*, she meant to say. *I am near to death. So is this elder, my mate.*

Later she fed Drummer, for he could not walk well and could not carry his own food. She gave him some dried deer meat, some greens from the creek, and a leather pouch filled with nuts they had cached nearby the year before. "This day will be warm and long," she told him. "It is the warm season, and the day will be long."

He could see that for himself, but it was important for her to speak it, for speaking made it so. "I remember others like me," he said. "Yes, others." His speech was hesitant and slurred, but she could understand.

Still, she didn't reply to this, for he said it often, and there was no reply. Instead she changed the subject. "It's a fine day to wade in the salty water and search for shellfish. A fine day to throw the nets we've been making through the warming season over the water and bring out the fish. When we have done this, I will bring you some."

But he would not be put off. "I remember others like me," he repeated. "Oh, no, not with my leg, not like that, but short, like me, and slow, but not as slow as me."

"They're gone," she said slowly. "We have seen none like you, not in a long while, not since my mother's mother's day. The day you came to us, Drummer, I was young. I had a child then. Crane. You hobbled into our camp. You were hungry, so hungry. You didn't speak." She paused, lost in reminiscence. "There was no one else. No one came after you." She trailed off. After all, this too they had said before, many times, once he began to speak.

"But I remember them." His thick voice was uncertain. She thought he repeated this phrase over the years so he would remember them: saying made it so. "A woman," he said. "A man. Others."

She stood and brushed invisible dust from her wrapping of light deer hide. "Only you, Drummer. No one else." She walked away.

He knew there was someone because he could see inside his mind. He was small. It was raining hard just outside, a broad, flat sheet of rain across the shelter entrance, hiding the valley, the opposite hillside. It was raining, but he was dry. Fire kept them warm. The men all seemed so big, but now he knew they were short, as he was, with eyes set deep beneath thick brows.

His people paid little attention to him back then; he was under foot. There were two other children with the group, but they were older. A woman fed him. She wasn't tender; he didn't remember her with affection, but she gave him food, and that was enough. There was not much affection among his people, not like the People here.

That was the night of the great rains. Fire flared up. It was growing dark. The sheet of rain reflected firelight in glistening red streaks.

Suddenly everyone was shouting. The men grabbed their spears and ran into the rain. The women followed, disappeared into the darkness, and he was alone.

He waited. He remembered waiting. He was patient, always had been, but he was hungry, and after waiting he looked for food. He found some, too, and pools of water left over from the rain. Days passed; he didn't know how many.

He dreamed through the day. This was called remembering. He was very small. The fires went out and he was cold. There was no more food. It came to him that he should follow his people, and so he set forth from the shelter. There was a stream and he followed it down. There was plenty of water, but no food. Walking was slow for him. The others, they had been much faster. His leg hurt sometimes and he stopped to rest.

Once he saw the great hulk of a rhinoceros grazing in a small glade beside the stream. He didn't remember its name; he supposed he didn't know because he was small. It had great shoulders covered with long hair. Since joining the People he had seen many, and always fear opened in his belly when he saw one.

Then, though, in his memory, he was about to ask the beast if it had seen the others and something stopped him. The beast glared at him with tiny eyes, long strings of grass hanging from its mouth.

He became afraid and understood that it would not tell him about the others, that it would hurt him. He crossed to the other side of the stream and continued.

The old woman touched his shoulder and he roused from his reverie. She handed him some smoked fish. He ate, grateful as always.

"The warm periods are shorter than when I was young," she told him. "The cold cuts deeper." This too she had said many times.

"The old always say things were better when they were young." He pulled a fish bone from his mouth and probed it between his large, square teeth.

She nodded several times, thinking it over. "Perhaps it's true." She was distracted, looking at him, thinking of Seer's words: *I did not see all of us there.* Drummer moved more slowly this year than last. His gait was more awkward, the way south through the mountains was hard, and the mountains were bitter cold this time of year even under the sun.

She and Drummer were seated on the grassy bank of the estuary. Sunlight dappled on the water. Long-legged cranes dotted the far shore, their necks jerking forward and back. Drummer's twisted foot trailed in the water.

The old man came and sat beside her. "How's my son?" he asked, looking at Drummer with a lopsided grin. The old woman knew this grin could not conceal his fear, his grief.

Of course he was not one of Drummer's fathers. When Drummer came to them he had a name. Besides, the old man was already one of Crane's fathers, perhaps even his sire. It didn't really matter; he was a good hunter and provided for her and the two boys.

Most of all, though, he was the one of the group who knew what was becoming manifest in the world and could speak the truth of it. His words, after the long night of whistling and drumming, the longest night in the cycle of warm and cold, were always true.

When Drummer appeared and she and Seer took him in as one of their own sons, the old man became one of Drummer's fathers as well. He treated both his boys the same. Already, before Drummer, he had been teaching Crane the ways of the spirits. Drummer pounded on things since he could hold a stick, so the old man apprenticed him as well. The four of them could summon all the spirits, the animals, air and water, the white light spirit from the mist and night. So Drummer became his as much as Crane.

She touched the old man's arm fondly, knowing these things, and seeing the sadness in his smile.

After Drummer appeared, not all Crane's fathers had felt the same way about him. Many spurned the newcomer as an awkward misfit, both strange and a stranger. One of the fathers suggested roughly they leave him in the forest, they didn't want him; he would only slow them down, so he said.

Besides this old man sitting here, only one of the other fathers still traveled with the band, but he was out hunting most of the time and seldom around. The old woman was pleased the others had drifted away to other groups and had not reappeared. Supposedly one had gone far north to the ice itself, but this she did not believe. He was a lazy man who liked fire warmth and ripe fruit too much.

They started south. On the way it came to pass as the old man had foreseen. Though the old woman protected Drummer as best she could and helped him along as they made their way to the other sea, he slipped on a mountain trail, sending loose stones flying from under his awkward foot.

With a startled cry, he tumbled from the path and disappeared into the mist below. They looked down into it for a long time before moving on.

As far as we know today, the Neanderthals were gone by around forty thousand years ago, leaving modern humans as the last hominin standing. We do not possess the wisdom or vision to see how long we will continue to endure, but from this point history belongs entirely to the human species.

Larger bands encouraged more complex roles and more specialized skills. In Europe these changes appeared on cave walls as painted or carved memories of the natural world given physical form. Such images were of no immediately recognizable use: they could not physically protect. Though they represented animals, they could not be dried or cooked as food. They could not provide furs or skins. Their precise purpose may be forever denied us, but surely their utility was nonmaterial, what we call spiritual. They invoke animals for hunting, protection, guidance, or healing. We call such images art, a word perhaps as ambiguous and vague as religion.

As with the Neanderthals, most artistic expression would have been evanescent: sounds, rhythms, characters, ideas, and representations of the world and all it contains. Music, dance, story, and song leave the very faintest of traces, a 40,000-year-old bone flute from Germany, for example, the oldest known. Other flutes of mammoth ivory, wing bones of a mute crane, or a bone from a Griffon Vulture prove the ubiquity and endurance of music.

Singing and rhythm are so basic to humans that one theory suggests language itself evolved from Neanderthal song. Whatever its form, performance is our way of meeting an evolutionary need to understand the world and our place among the visible and invisible forces that animate it.

In Darkness

By twenty-five thousand years ago world population was perhaps three million. With more people came more questions about the world and more ways to answer them.

The continent-wide tradition of cave painting suggests another source of answers: specialists in perceiving and navigating an unseen world and oracles of the future. We call them shamans, from a Tungus word for a priest of the Ural-Altaic people. They worked by altering consciousness with drumming and song and fatigue.

In many subsistence cultures today, shamans are the odd ones, the lonely ones who, by their separateness have access to a world of spirits others cannot see. They are the ones who, by their special, separate natures can fly, interpret intents and actions, and cure disease.

Shamans were effective, too: their stories, passed down from a clouded past, helped their bands thrive in an unpredictable and bewildering world.

The camp sprawled across the wide meadow. Children were busy play-acting, men sat in small groups working flint tools, and women sewed clothing.

One stood apart, watching. He, if it was a he, was called Changsar. He may have been a man, but many said he was not a man at all. They said he was a spirit who could take many forms, or, like now, be invisible. When he was invisible, he did not exist. Sometimes he was a woman. Other times no one could say exactly what he was, or if he was.

Most of the time Changsar considered himself a man, part of the band, but not part of it at the same time. Whatever he was, he had been carefully watching the boy named Farl all through the warm season. He knew as well as anyone the spells of weakness that had plagued the boy from the day he separated, and how the boy's mother would refuse to leave him, how she always found a man willing to carry him.

Like others in the growing band, Changsar had often heard the boy screaming from night visits and tell in a trembling voice what he had seen in the world of dreams. Changsar had also seen how the others sometimes looked away when Farl appeared, how they managed not to see him.

Just now several of the younger named ones were animals in front of the shelters, a herd of red deer grazing, or lifting their heads, alert to danger, or bending to crop the grass. Some wore skins or had fashioned antlers out of sticks. Clouds scudded overhead, splattering the camp with intermittent sunlight.

At the edge of the group Farl stalked on all fours, head low, neck arched, mouth open, upper lip curled. He was following one of the girls crouching low before him. She was staring into his shining black eyes and struggling to suppress her giggles.

Farl's eyes locked hers. He shook his head, inhaled loudly through his teeth and blew his breath out through his nose in short bursts.

The girl, who had only nine full cycles, squealed. She backed away, turned, and ran, looking back, expecting him to follow. But he did not follow, he stayed where he was, face distorted with disappointment, longing, or rage.

Changsar was certain now that Farl was the one. It was late already in the warm season, but the boy had grown and the change was on him. He would not play at mating much longer; he would be nearly complete and able to mate as a human, not merely imitate a four-foot, a plant-eater stalking a female of his kind.

Before that happened, he would have to take him.

The band had grown since the days of the Fathers' Fathers, and today numbered twenty hands of fingers, a very large number even to Changsar, who had seen many seasons. Farl was the oldest of the named children among all the families of the People of Grass.

Farl seldom spoke, and the other young ones watched him always. They were a little afraid, for he was strange—sickly and unpredictable and given to frightening visions. In spite of this, or because of it, they often copied him, often did as he did, and today he was a stag in rut looking for a doe. Already several of the other boys were lowering their heads and curling their upper lips.

The adults were amused at these antics. Look, they would say, pointing, Farl is the red deer stag. Did you see how he followed the doe, breathing in her scent, the scent of her urine, the scent of her scat! He wants to know, is she ready? Yet how could he know of such things? No one had told him. Like the other young he was seldom away from camp, seldom with the men when they hunted; even though young and inexperienced he knew the actions of animals, he knew their minds.

Changsar could wait no longer. Only once in his life—just after the wise one who taught him had left his body, in a time when the band was much smaller—had a boy like Farl been born among the People of Grass. That time it had ended badly, and Changsar had gone away for a long time. This time, he told himself, would be different. This time he had a life of experience.

The girl jumped to her feet and trotted to the fire, drawn by the smell of roasting meat, and the game was over, just like that.

Changsar took this opportunity to make a sound. Farl looked around. The man-shape gestured, fingers down, for the boy to come.

Farl slowly climbed to his feet, but once standing he stood still, dazed.

The figure gestured more urgently, and the boy, lips still curled, walked toward him stiff-legged. When he was close, Changsar spoke in the softest voice, the most calming, most persuasive voice. "The doe. You seek to know if she is ready to mate."

"Is that what I do?" Farl wondered, still dazed.

"Do you remember this?"

The boy tossed his head. "It's something I know. I can't say."

Changsar tugged thoughtfully on one fork of his sooty gray beard, for now he was a man of flesh. "You will come with me."

Farl stared.

"It is as the Fathers say, as the Fathers speak. This is not for you to decide. You have a name, but you are not yet a person."

The boy lowered his head in agreement and followed Changsar into the forest without a word.

They walked through the day. Farl wanted to ask where they were going, this Changsar could clearly see, for at one time he had been such a boy, following. But Farl said nothing. Clouds drifted, gray and heavy, overhead. The light dimmed, and still they walked, old and young. When it was on the edge of night they stopped. Changsar handed Farl some dried meat and they sat side by side in the darkness, chewing.

"Where are we going?" the boy finally asked.

The old one didn't answer; he was asleep.

The next day they climbed low hills, then up steep slopes of scree and shrub. They didn't stop until they reached a brow of furrowed gray stone over a dark opening. Seeing this, a fear passed over the boy's face for the first time. "I have seen this," he said softly, staring into the dark abyss. "I know this."

"Of course, in the dream world, the sleep world," Changsar said in a way that stopped all questions.

"I didn't like it," the boy offered in the same tentative voice.

"You will like this far less." Changsar seized the boy's arm and dragged him roughly into the darkness.

They stumbled over slick stone and rough stone. They brushed against spiders' webs, and against smooth wall. They stepped up and they stepped down, turned first one way, then the other. Farl knew they would never leave this place, no matter how familiar it was to him from his visits to the dream world, from which he sat up so often whimpering in terror. And now, when he tried, he could not rouse himself from the dream.

Changsar had spoken not a word, had made not a sound. He stopped the boy with a hand held back against his chest. After a time in black silence, fire starter flared to flame and flamed to torch. In fragments, Farl saw dark brows, wild hair wound around with scattered scraps of fur, feathers, and bricks of dried dirt. He saw the lines and pouches of an old man's face, a scraggly beard standing away in tufts, and his fears redoubled and set his legs to trembling, for he did not know what this being was.

"Sit," Changsar ordered, and his voice was dark and rough and unkind.

The boy sat. The ground was clammy and he could see nothing outside the tiny circle of light flickering below that awful face, now leaning toward him haloed with utter darkness. He jerked back and hit his head against the rock. He was far underground, his mouth stopped with dirt, or so he felt.

Changsar's eyes bulged, expanding outward. He leaned closer, lifted his open hand palm up, and blew something into Farl's face, something that startled him into sucking in a breath.

That was when his harrowing truly began.

The old one must have set down the torch, or lamp, for it stopped moving. Changsar, or whoever this was, held a hand against a wall the color of dead fish belly. He blew again, spreading a red stain over the wall, and when he took the hand away (when did he do that?) a hand remained. The old man had grown an extra one.

"Look, boy!" Changsar's voice had changed again, deepened, slowed, as though smeared with animal fat; Farl could smell burning. He heard rustling at the edge of darkness, and then the darkness was complete.

He could not know how long it lasted: longer than a night, longer than a moon cycle of nights. Sounds—sniffs and grunts, snarls and moans, thumps and rustles—came and went. Then they were rhythmic, steady, a pulse. The pulse went on. Farl could not endure it. He forgot his name, his place, his people. He was no one, a mote swirling in a tiny eddy in the current of a dark waterless river. He was aware that he was hungry, thirsty. Then he was not.

Gradually he recognized the rhythm: a child playing with sticks, hitting them against something hard, stone or shell or wood, he couldn't say. A child's game, before naming, or after.

The sound stopped and silence choked him. He had nothing left but fear, out of which Changsar appeared as from heavy rain, sleet. "The ghost hunters come," he intoned, voice hollow in the thick, starless space. Farl felt heavy stone over his head, ready to fall if he moved. "The dead, they come," the intolerable voice insisted. "They come for you. They will take you to their land, which lies beyond the stone. You will enter the stone. There is no air."

Light flared, orange anger against a white wall. The hand outlined in blood waved at him. He drew back. He didn't mean to: his body did it. Fear did it.

The voice said, "That is a person on the wall. You are not a person."

"Not ..."

"Not a man, but a deer. Look!" A hand passed through the light, and the light moved, and the skull of an enormous cave bear stared at him with empty eyes. The light flickered over teeth, over smooth bone, over sloped forehead. The bear was alive; it was looking at him. And then the voice continued, "You hear him coming though. You hear him."

"I ..."

"He carries a spear. He is many. Many spears, sharp points coming. You are afraid."

"Y-yes."

The voice continued. "Man is coming to you. Man, hunter, dead man, ghost man, he must eat. You know the animals, the way they curl their lips, the way they smell the female to know if she is ready for him to mount her. You know this."

Farl grasped at this. "I've seen ... ."

"Be still! To know this is not to know death. You must know death. If you do not know death you do not know the animal, the deer, the fright of the deer, the panic, the run, the exhaustion, the lolling tongue, the struggle for air. You do not know the end of the deer. You must know death!" the voice screamed.

Drumming started again, soft and slow at first. Farl could see only the wall, the awful white ripple of wall, with the dreadful white hand outlined in blood. The beats came, slowly, from far away, but they came closer, faster now, and the voice said, "You are the animal, the aurochs, the bison, the deer, the horse. Human comes with his spear. He hunts, she is hunting; they hunt you. Spear flies from darkness, from grass, from air. It touches your side, your buttock, your chest, plunges into your belly. You, aurochs, bison, deer, you will die so humans may live. You must feel the spear; you must know this death, feel the life draining away, fading. How can you be the living animal if you cannot know its death?"

Out of the darkness Changsar's hand, the real one, appeared. It held a stick of burned wood, and black soot lines sketched quickly on the dead white wall: legs, buttocks, part of a person, that was all, just part. No face, no arms, only a man trying to run.

"You see." The voice was far away, warbling as if disrupted by water flowing. "You, child, you, great beast, you deer, you bison, you horse, you." The hand struck hard at the wall and a line stopped in the animal buttock, the human's buttock, the deer, the horse, and terrible pain seized Farl's body.

"You felt it," the voice stated at the same time, and Farl tried to scream away such searing pain in his buttock. Blood flowed. Before he could scream again another spear struck his back. He clawed at the heavy spear hanging from his ribs. A third struck his upper leg and this time he fell, his life pumping out with frightening speed onto the broken ripples of stone dyed red by flame. Farl saw the spears hanging from the body on the wall; he felt the stone points deep in his own.

The pain was terrible. He could not endure, and so he died.

As he was dying, his penis rose; he felt it strain. This frightened him more than the spears, more than death. Never had he felt such fear, such piercing terror, not when he watched the horse mount the mare, the bison mount the cow, the rhinoceros pawing at the female's back and the thrusting raw need of it. The fear was a thing, a monstrous beast, larger than a mammoth, stronger than a rhino, more implacable than the great ice he had seen only once when he was not yet named.

He felt his flowing pulse, the release. The light went out. Changsar vanished, and Farl was no more.

The darkness was absolute, though the idea of absolute was strange. The silence also was absolute. He listened to the silent drum, looked at the black emptiness, and only after a very long time knew that the drum was his heart. He was alone.

But he was also Farl, a boy of the People of Grass. No, he remembered, he was not a boy. He was grown. The memory came like one of the small birds that fluttered around the camp in the early warming season after the worst of the cold. It was there, it jumped, and was somewhere else.

It vanished behind leaves, into the crowns of trees, hopped to a branch, from a branch. It cocked its head and looked at him with a wise, bright black eye, and he knew it really saw *him*, as he was, naked and alone.

He stood in the darkness. After a moment of puzzled thought he made a sound. He listened to the sound, expecting it to fall away, but it did not die. He felt the chamber around him and knew that on the wall behind him were lines that were part of a person, male or female, he could not say, struck by spears. That to his left there was an opening, and above him a great dome, and to his right the wall curved close. All this he knew.

And then he knew that he felt no pain where the spears had struck him. He felt no wounds on his body, if he had a body. Of this he was not sure. It was of no importance whether he had a body or not.

He turned to his left and began to walk, hesitantly at first, and then with greater confidence, humming as he went. In time he emerged from under a brow of furrowed stone into the light of a sun just rising.

Two days passed before he met again with the People of Grass. He greeted them, but they didn't recognize him, and when he asked after Changsar they looked at him and said, "Who? There is no such person here."

For the most part, we no longer see spirits. Our ancestors' understanding was imperfect: they practiced magic and superstition, to our eyes, though not to theirs, a waste of time.

In *Faces in the Clouds: A New Theory of Religion*, anthropologist Stewart Guthrie describes how evolution drove us toward seeing agency in inanimate things that raised the hair on our necks. How we responded to an ambiguous sound or sight could save lives. The spirits may be gone from the world, but the terror remains. We deliberately recreate it in books and horror films, but in the deep past if the sound was a tiger and not the wind, the cost could be high.

Humans developed a sense of what's needed to live in groups, particularly emotional commitments to one another. And so they tell stories to explain and soothe anxieties about existence and mortality.

People drawn to ecstatic or mystical experiences have a desire to pass them to the next generation. Herein are the beginnings of a sense of the past, of tradition, of lineage.

Shamans applied, and in many cultures still apply, a profound understanding of natural history to human development. The transition out of childhood, as any parent knows, can be chaotic, painful, and sometimes tragic. To facilitate, the shaman could scare an adolescent into adulthood. If done right, initiation ceremonies, especially for boys, are life-altering experiences. The shaman exercised what prehistorian Jean Clottes called "shamanic flow," individual, ad hoc, and spontaneous access to hidden insights.

Changsar, and Farl after him, had a vital role in the group. Placebo effect or coincidence may account for cures. Intuition achieved through practice may convince people to do spectacular things for their own survival and that of their peers. A gifted shaman could perceive invisible currents of desire and fear and direct them toward enhancing life. They were psychologists above all.

Farl's initiation was a unique event, improvised, interactive, and personal, never repeated. Deprivation, repetition, fear, and readily available (and carefully studied) mind-altering substances enhanced the experience.

The general outline is common in preliterate societies. New forms built on old ones were constantly reinvented as needed.

Terrifying adolescent boys created a bond among all who endured such vivid, unforgettable experiences and kept them engaged in their band, with empathy for those who suffered and control over their own fear and aggression.

Remnants of such initiation practices still exist in the bar mitzvah or first communion, but for the most part they have gradually sunk beneath the veneer of civilization.

Boy and Leaf

Climate shifted from cold and dry to warm and wet and back again many times. There came a time when the ice began to loosen its hold. The land opened up and allowed people to flow down into the innumerable valleys cradled in the mountains to the south.

Daily life was always about the immediate present. The big human brain's survival depended on continuously inventing new ways to master the world.

Sometimes that mastery was expressed in paint.

A heat-fever during his separation took his birthmother before his name had found him, so Boy was the name he had. No one told him he should mind it, so he didn't. It was a good name.

There had been a child named Girl too, but she met someone she liked with one of the southern bands two winters ago and they hadn't seen her since.

Boy's mother had left her body for the Unseen when she and Boy separated. Several women witnessed it. Like a sudden wind on calm water, they said. Boy separated in a spray of blood. One of his aunts had told him this. A spray of blood, crimson like hazelnut leaves in autumn sunlight, like ocher.

Late one night when it was still dark with a sliver of moon and he had twenty full years already, Boy was listening to the women, their constant chatter. The buzz of their talk barely paused whenever a wail of pain rose up and twisted amid their high gutturals.

The women, old and young, were gathered around a motherbaby. Whenever a woman felt the separation coming on they gathered, as now, hiding her from the eyes of men.

First light filled the entrance and the women started singing to ease the separation, low humming at first, one woman, then another, all the aunts and sisters joining, stroking the motherbaby, coaxing out the new one.

The men were waking, rising, stretching. One by one they wandered off into the light, going down to the river or away toward the fragrant trees. Soon only Boy remained, leaning on one elbow, back against the rock wall, listening to the swelling melodies, the rhythmic clapping, his eyes closed. He was living his own separation, the shower of blood.

This time though there was no shower, no flight. The singing stopped, the baby cried, and he could hear the suckling. A girl, one of the women said.

Was he the father, Boy wondered? Perhaps, but it wasn't important. He had lain with the woman, but so had others, Flint certainly, and, he thought, that man with the scar they met last year when the leaves were turning, the one from the eastern valley beyond the mountains, the skilled carver they knew as Minz.

Boy's band will raise the girl, as it raises all newcomers.

One of the women moved away from the others, the one called Leaf. She was a little older than Boy. He watched her approach.

"You're going?" she asked.

He stood and brushed himself off. "How did you know?" He smiled, acknowledging her interest. "Ah, you feel it, the need."

She dipped her head and her long hair fell alongside her cheeks, shadowing her eyes. "It's our last day here. Tomorrow we move to the sea. You know this. I know this. So, today ..." She hesitated. "Today we go. We must."

"Yes," he agreed, collecting his things. He nodded toward the women examining a long net. "We'll have rabbit tonight, when we return," he suggested.

"Not many rabbits left here," she replied. "Some of us yearn for the sea."

He slung his pouch over his shoulder. "We must go."

They climbed the mountain above the shelter in silence, pausing only to look across the valley at the opposite mountainside. A small herd of red deer dotted the grass below the timberline, drifting slowly up into the scattered trees.

"Today or tomorrow they'll go through the pass to the next valley," he murmured. "They will feed the Sunrise band, then. Come, we must perform the summoning, you and I."

They reached the mouth of the cave, tucked against a corner in the cliff face, by a narrow path. The drop to their left was nearly straight, a long way down, halfway to the shelter, maybe, but neither of them glanced that way.

They leaned into the opening. The small antechamber, dimly lit from the outside, was empty. They both sniffed, turning this way and that. He entered and sniffed at the opening on the far side, just a few steps away. He looked at her. She shook her head. "Nothing," she said. "No bear."

He grunted and crouched. He took out his fire pouch and moments later had blown a spark of dried fire starter mushroom to life, transferred fire to his small tallow lamp, put the embers back to sleep and returned them to the pouch. The hollowed stone glowed. He picked it up and set out without looking back.

Out of sight of the entrance, chill darkness settled on them. The air did not move, and when he stopped to examine the wall, the lamp flame settled into a stillness cupped around their breathing, the muted thud of their hearts, the smell of stone and water, and a faint steady drip to their left.

He followed the sound and the motion of their passage made the light jump and sway until the uneven rock walls and ceiling seemed alive. In places they had to squeeze through narrow openings. Once the lamp went out and he had to light it again. Finally he stopped at the end of a short straight passage, inhaled slowly, let the breath go. He held his palms over the wall, moving the lamp from side to side. "You see them?"

"Yes," Leaf said. "The curve of the rump, the neck. Deer, does, yes, three of them." She went on, "But look over here. One on this side, going toward the entrance." She traced an outline with her hand, not touching the stone, and a deer leaped into view, her head turned slightly toward them as she ran.

"Yes," he murmured, "this is the place."

He unrolled a leather apron on the damp floor. The space was narrow, just enough room for his pouches of red ocher, an unused block of it, a few swabs of leather, a small grindstone. He moistened his fingertip and touched the red powder. With careful dots he outlined the deer, running away from the entrance into the darkness.

She watched him for a few moments. When she turned away toward the opposite wall, she saw a young doe, leaping forward toward the distant light.

These were not large, these deer, just a hand or two high, thickly outlined in red. Boy and Leaf had seen larger animals only a year gone, farther to the west, in another cave, a larger cave, animals with shading and detail, with a sense of depth and distance. They didn't know who had painted those animals; they had always been there. Those animals were nearly life size, and very strong. In the second space where the cave opened out someone had seen and summoned into view two bulls, tails and heads down, upper lips curled, sniffing. It was their season and the bulls were looking for viable cows.

These, the ones coming into being under their fingertips, were the ones they had seen outside across the valley. They were the same size, seen from a distance. They didn't bother with detail, placing only a dot for an eye, a line for a tail or a leg, yet these deer were as robust, breathing and alive as any.

They tapped the dots onto the wall and smeared them into lines. Neither hesitated, neither stopped to correct. They had drawn all their lives, in charcoal on stones along the river, on dried bones found in the open spaces, in ocher or black mineral and water on leaves and birch bark and scraps of leather.

When they had finished, they looked at their work. Four deer flanked the narrow space twisting away from them. "This place reminds me of the motherbaby that separated this morning," Leaf said, indicating the bend further back. She started tipping her red powder back into Boy's pouch. "Narrow. We had to turn her like this, to help the separation."

"My separation killed my mother," Boy told her. "She was called Silence, and she left."

Leaf nodded. "I heard. There was much blood. Once a motherbaby died and Flint cut her open, do you remember? We could see how the young's feet were caught."

"This morning?"

She sniffed. "This morning was easy. The newcomer was turned back as she should have been and Amber caught her, just right. It was a good morning, a good separation." She paused. "They aren't all so good."

He grunted agreement. "There's something wrong with your doe."

She shook her head and her dark hair floated. "No, come." She took his hand. "Bring the lamp."

At the far end of the narrow passage they turned back. Faint daylight from the entrance added to the flickering lamplight and the doe leapt into life, running toward them, perfect. They could see her breath, for it was autumn and something was behind her, out of sight around the curve of the wall. Boy was sure it was a buck.

"Yes." He nodded slowly. "Yes. They will mate."

At the entrance he said, "Wait."

She waited.

How often they had done this, she thought! How often they had climbed to a cave or shelter and brought forth the animals! And how easily the animals slipped through the stone, as easily as through the air! She watched the deer on the opposite slope while he took the red powder into his mouth.

He placed their palms side by side against the face of the cliff beside the entrance and blew a spray of red over their hands. When they took them away the hands remained on the wall, still touching it. He rinsed his mouth from his skin of water, packed his things, and they scrambled back down toward the smoke. "They've caught rabbits," she observed with a smile.

When Boy and Leaf reached the shelter, the girl child and her mother were sleeping. Most of the men were away foraging for food or stone. Only one old man remained to tend the fire. He lifted a heated stone between two sticks, waved it at them, and dropped it with a hiss into a cooking pouch. He added some greens collected from the hillside and looked up. "Rabbit stew," he said with a grin. "No deer today."

Boy nodded. The band would eat well. Tomorrow they would leave for the coast and their diet would change again. That was always a good time. Next year the deer would be abundant in this valley. He and Leaf had seen to it.

The women had gathered on one side of the shelter with the recently separated motherbaby. The quality of their chatter was different, no longer the lighthearted polyphony of the early morning's easy separation, but the low, somber tones of concern, of uncertainty.

"Doesn't sound good," Boy said, shaking his empty water skin.

"No," Leaf agreed. "The newcomer is weak, and they must decide. They will ask old Tesk to help, I think."

Boy's brows creased. "It's that serious?"

"It may not survive the walk to the coast."

"Nothing to do, then."

When they returned from the river with full water skins the women were seated in a circle on one side of the shelter. Old Tesk was cross-legged in the middle, rocking her head back and forth, sucking air over her few remaining teeth. She was an elder, with more years than could be counted on the fingers five times over. Everyone knew she had an opening in her mind to the world just out of sight. "Eee, eee," she keened, her eyes closed. "Eee, eee." She began to speak without words. This went on for a time. Then she asked a question and stopped, looking around, and her eyes, normally filmed and dull, were bright.

No one answered, for the question was not for anyone here.

She shook her head and her sticky gray hair barely moved.

Across from her the new mother held the baby on her lap. She was looking at old Tesk, though from time to time she glanced down at the newcomer. It was asleep, fingers in its mouth. The sun crossed overhead and still the wordless questions went on without answers.

Boy pretended to be busy, but really he was curious. From the corner of his eye he watched Leaf settle next to the new mother. From here, on the other side of the shelter, it looked fine to him, but he knew little of such matters. This was something for the women to decide.

Tesk sipped from a small wooden vial, swallowed noisily, and spat on the ground. After a few moments her eyes rolled up and her head fell back. The women began clapping a rhythm and gradually speeding up until the sound was almost continuous. This went on for a long time. Suddenly Tesk fell over with a shout. Leaf caught her and lowered her to the ground.

One of the other women took the baby and carried it out of sight into the trees. Some time later she returned.

"A name will never find it," Leaf told Boy later over rabbit stew.

Already the women were chatting, at ease. Even the new mother seemed to have forgotten her newcomer.

The cave painters were superb observers and gifted recorders of their immediate physical world, its folds and turns, its fears and delights. With simple lines they captured *from memory* the essence of animal mating, hunting, grazing, or stampeding. Their understanding of animal anatomy and psychology was profound.

The doe Leaf drew still runs toward the opening of a cave called Covalanas in Spain, eager to reach the light. Many other deer sketched freehand by a dabbing forefinger dipped in red paint share the same dark, cool space. Though simple outlines, they are accurate in proportion and very much alive.

Leaf's doe is unusual, though, stretched and distorted when seen from the point of view of the painter but perfectly proportioned when viewed from the cave entrance. Her fingertip felt her subject through the stone and evoked it one dot at a time. She could visualize it both as she was drawing it and as it would appear at an angle from a distance. Only someone whose eye had not been constrained from childhood by the straight lines and right angles of today's world could have translated so fluidly from one dimension to another. There is a similar example at Lascaux. This painting technique, called anamorphism, was lost until reinvented by Leonardo da Vinci.

Cave paintings have provoked many theories of their meaning: art for art's sake, hunting magic, instructional illustrations of animal anatomy. Since they are found in nearly inaccessible chambers where the only lighting would be from torches or small oil or animal fat lamps, access was difficult and frightening. The spaces were often claustrophobic, with dead air and a sense of oppressive weight.

Flickering light lent motion to the images. These were the first animated films, magical at the time. They were exercises in visual memory that captured the motion of a horse's flowing mane and flaring nostrils; they evoked the animal itself through the stone and maintained communion between human and animal. Their anatomical and behavioral accuracy make them instructional, certainly, but when created in the company of others in such difficult locations they must have had deeper meanings, more transient and performative, for they were, above all, performances. We know this because they were often abandoned or painted over.

The caves, often blocked by rockfall, preserved the paintings in still silence. Now that they are open, alas, the paintings are rapidly deteriorating.

As climate warmed and larger animals, both predators and prey, grew scarce, the human diet turned to smaller creatures like turtles and rabbits.

Cultural evolution speeded up. Social norms were changing ever more rapidly. New behaviors around sickness and death and more complex family structures emerged like new species.

The Snail Creek Shaman

A lush green world followed the Ice. The Holocene arrived in fits and starts, with lengthy setbacks and sudden leaps, but the impacts of a warming climate on human life were consequential.

People could stop wandering in search of food and settle down. Among the first lived in the Levant. We call them Natufians (14,500–11,500 years ago). They built clusters of small round wattle, mud, and stone houses where they could linger near their food year-round. Their understanding of the ways of the plant and animal world took new, unexpected directions, among them the first tentative farms.

Old traditions confronted these new realities. Around twelve thousand years ago a woman was buried in a cave in Israel called Hilazon Tachtit, Lower Snail Creek.

She is the Snail Creek Shaman, and her grave tells a story.

They toiled in single file: old men, old women, youths, and children. Near the middle of the line two men newly come to breeding age dragged an old woman lying on an improvised sled. She was twisted awkwardly to one side as though in pain. Her thin hair, cut rough and streaked with ashes, fanned out from her head. She spoke softly to herself as they dragged her up the uneven slope.

From time to time the young men had to hoist her over a particularly rough part or rocky outcropping. Progress was slow.

The winter had been colder than usual and there had been no rain at all since then, but this afternoon was warm and they were sweating, the two burdened ones most of all. The only sounds were grunts of effort, the scraping of leather sandals on the dusty slope, and the swish of sand and gravel running down from each labored step. There were twenty-three people altogether, including the bearers.

Two of the older men carried baskets of living turtles, dozens of them. One paused to lift the lid and whisper, "Go to sleep. It's hard enough carrying this basket without all that clambering about."

The others laughed. "They know what's going to happen," one of the elders said. "That's why they're restless. You'd be, too, you know."

The old woman's niece came last. She was almost old enough to be a person, and if she decided not to become heir to her aunt she would be ready to find a man from another group. Almost everyone believed she would choose this path and leave the settlement, for she had shown little aptitude for the old woman's difficult ways.

Slung over her shoulder was a sack of wild goat hair dyed a rich golden color, containing the old one's eagle wing, most of a leopard's pelvic bones, the core of a male antelope's horn, and two new marten skulls still covered in soft, clinging fur. The girl knew these contents well, for she had prepared the sack herself under her aunt's demanding eye. The old woman may not see well any more, but she *knew*, somehow, everything that happened in the group.

Everyone knew the sack was not heavy and the girl had no cause for complaint, but she was so visibly unhappy they were glad she was at the end of the line.

The leader at the head of the line stopped where the slope flattened for a few paces. "We rest here."

The one who whispered to turtles said with a grin, "You are wise, as always."

She laughed at this. "Not wise. Old is what you mean. But I'm not so old as you, nor so foolish."

"Mind who you call foolish," he answered, laughing with her.

They sprawled gratefully on the stony slope, looking toward the endless waters to the west. This blue infinity bounded the side of their world where the sun died every night and became the ghost that floated, ever changing, through the embers of the day.

Their pause was brief, though, for the old woman was not dead, not yet, as she reminded her bearers often on the climb. She pushed herself onto a bony elbow and looked at that distant band of blue through watery eyes. But after too short a time she waved her hand so the loose flesh under her upper arm waggled. "Enough!" she cackled. "Already afternoon. Much to do! My time is soon on me. On me," she repeated softly. "Go, then, up. Come on, hurry."

They collected their things and started on toward the place she indicated, an oblong shadow in the limestone escarpment near the top of the cliff still some distance above.

At last they reached a series of shallow ledges covered with the spiky stalks of dried shrubs and a few hardy wildflowers. "Put me down," she commanded, and they did as she ordered, for she was the one they called *Zuen hildkoen-hitz*, She Who Speaks with the Dead.

The group needed her more than ever. No one remembered when the hard times began; it seemed it was always thus, back through the mothers, and the mothers' mothers, hard winters with too little to eat, and hot summers too dry for comfort. Scarce animals and plants too widely scattered. Though she knew stories of a different time, a warmer time when the plain below teemed with antelope and the hillsides were bent over with food, she told these stories less and less often and some time ago had stopped entirely. Mostly, because she could tell them where to find enough to stay alive and cling to their village, they followed her advice.

And now she was leaving them. Though all knew this, they pushed away the anxiety of the time the *Zuen hildkoen-hitz* would be gone. Her niece was still too young, and too sullen, too *normal*, perhaps, and so unlike her aunt. When the old woman was gone, who would tell them how to live?

Down the slope the thin ribbon of water that wound through the valley had shrunk to a bright thread. Far to the east was their home settlement, the cluster of huts they left every day, the place they came back to at night.

Some, but not all, avoided looking that way, for they did not want to be reminded why they were scrambling up this cliff.

"Open, open," she croaked, showing her few worn teeth. She indicated the mass of brush. "We must go inside."

Her bearers cut away the crackling shrubs along the escarpment with flint knives and revealed an opening in the rock the width of eight people lying toe to head.

One of the young men, wiping his brow with the back of his hand, turned to the old woman. "How did you know of this place, *Em'kume*? No one has spoken of it. There are no songs. *Zaharaita*, Old Father, said as much only last night." *Zaharaita* talked to the turtles.

"You ask me that?" the old woman hissed. "You might as well ask me how I sent away your night heat when you had just six years, Azkar!"

He was humbled. "Yes, *Em'kume*, I regret my question. If not for you I would have died then."

"Remember it, young one. This is knowledge you already have; you will need it again one day soon."

Already the turtle whisperer, who was called Old Father, was tipping embers from the hollow stem of a giant fennel and blowing them into flame. Others fed twigs and larger sticks to the growing fire.

Sun slanted down across the mouth of the cave, gradually bringing it into the light. The old woman showed her few remaining teeth in a half smile.

"You will eat, *Em'kume*?" the oldest man asked.

"No," she snapped. "I will not eat. I will drink what I have brought with me, I will burn what I have brought with me, I will breathe in the smoke, I will say what I must say, and I will follow the sun. Now leave me in peace."

She was moving away from them, not in the world of bone and stone, but in spirit. Her limbs trembled and her head kept tipping only to snap up again. From time to time she sipped from a small skin, and each time her eyes grew more vacant and distant. The people had seen her move away like this before, but now something was different: the words she was chanting were indistinct, rhythmic, repetitious, rising and falling, swelling and fading. And as she drifted away from them, she became more difficult to see, more like water, or the mists that come with cold winter mornings.

Once she noticed them watching and grinned with obvious effort. "Eat, eat," she told them. This time she did not raise her arm in her dismissive way but produced a small package and gestured for her niece to burn it nearby. She leaned down to inhale the pungent smoke, and fell to chanting again to herself, or to presences only she could sense.

The two young men were patiently waiting. "Ah," she said at last, addressing the younger, whose eyes were bright and inquiring, unlike his companion, who stared off to the south without expression. "Why do you stop, Azkar? I told you, you must carve me a place to lie in the bones of the mountain. You will gather around me, you understand that, don't you? In there," her bony finger thrust at the dark opening, "is where I must send out my breath. I've been watching you, boy. You are fast, as your name means, but there is something else ... something else."

She resumed her mumbling, gone from them once more.

There was a depression in the floor not far from the entrance and there the younger men began to cut the stone. Their flint tools tapped, *tchuk, tchuk*, and a hollow widened into an oval. They were able to pry up a slab from time to time and set it to one side.

When it was as deep as the length of the old woman's arm, Azkar scrambled down to the creek far below and brought back a skin of water. He scraped dirt outside the cave mouth and slathered the inside of the freshly cut pit with mud. He and his companion carefully pressed stone slabs into the bottom and sides. Azkar did not know why they did this, and the other did not care. It was as their *Zuen hildkoen-hitz* had commanded. Because she had a special power, and could converse with the unseen, he listened carefully and did as she said. There was no one who would deny her.

The niece was busy laying out the contents of her sack on a blanket spread on the floor of the cave. Azkar sat cross-legged beside her and one by one touched the objects and murmured words. The girl looked at him in wonder that he should know their powers and what to say to them. When she asked him he shrugged. "I feel it. I've always felt it," he said simply. She did not press him. She didn't feel it, whatever it was, but was so afraid of her aunt that she would never say so.

They had gathered outside the cave mouth in sight of the old woman, and from time to time looked sideways at her while pretending not to, trapped in their somber mood.

Only when the mud had thoroughly dried to a hard surface did she look up, though she seemed disoriented and confused. Azkar touched her arm, and it twitched violently. "What?" she snapped.

He gestured toward the pit. "It is ready, *Em'kume*."

She squinted, cocked her head to one side, and blinked several times. "Did you line it with mud? It must be smooth."

The boy nodded, for he understood that her vision was failing. The other man, crouched by the hole in his stolid way, gestured with the cow scapula with which he had been smoothing the sides.

She nodded, working dry lips and biting down on the stub of a stray tooth. "Very well, very well, yes. Yes."

Joints of meat, including an oxtail and a boar's leg, were set out on wooden platters. The women were cracking open long bones with hammer stones and scooping the marrow onto another plate.

The men were carefully prying the bottoms of the turtles to get at the meat without breaking the shells. Soon the hand-sized animals, dozens of them, had joined the other meats sizzling on the fire. Meanwhile one of the women had dropped heated stones into a leather water sack for a vegetable stew.

There would be food enough for all, today, at least.

A light breeze swept wood smoke across the cave opening and the old woman, who could speak with the dead, grew even more indistinct, her eyes lost in distances no one else could see. Still, her faint smile suggested she might be enjoying the smell of the feast.

The group tried to ignore her, to pretend this was an ordinary celebration of spring. Only Azkar glanced at her and quickly looked away. He knew she was moving deeper into trance, for he had at times, seated at the edge of firelight or shadowed under a tree, followed her partway into the unseen world. If she had noticed his presence there she had never spoken to him of it.

When he saw patches the color and texture of beeswax appear at her temples, he was not surprised that she gestured him to her side. She wanted him to prop her against the wall of the cave so she could see what was going on. This he did, attending to her carefully while her niece looked on. Sitting up this way her useless leg was obvious, flopped sideways. All her life she had been dragging that thing around. Now she would never walk again.

"Something I need," she said loudly. When everyone was paying attention she continued in an even, monotonous tone. "I see the sky filled with living things. I see the earth. I see beneath the earth, and in the waters, thick with movement, so many, so many. There is much voyaging to do. Much walking. Pah! What's to be done?" She slapped her hand on the useless leg. Her head slumped and her shoulders folded inward.

Azkar, leaning forward to catch her words, was suddenly pale. Muscles in his jaw tightened, and his fingers clenched the edge of the blanket holding her tools: the leopard pelvis, so rare in the world, so precious; the marten skulls; the antelope horn.

The *Zuen hildkoen-hitz* was gone into the other world and did not see Azkar, not at all, but her spirit tore through him like a spear and he lost his breath for long beats of his racing heart. He was Azkar, the swift. Had always been.

He saw clearly what must come to be. The *Zuen hildkoen-hitz* must walk in the other world. She must be free to voyage, to make the necessary migrations, to follow the scents and signs.

It was the only way to save her people. His people. The people. The only way.

This was a fear such as he had never known, not on all the hunts, not on the day they trapped and killed the leopard for *Em'kume*, not the time his brother was gored by the crazy aurochs, not in that terrible winter when the people were so ill with hunger. Those were easy fears, gifts, even, for they came from outside, from others.

This was different. This came from inside, from that piercing shock her spirit had seared through his heart.

She knew. He saw she knew, her eyes glowing in the semidarkness of the cave. She was there, beside her grave, and she knew. Once, slowly, she nodded. He bowed his head for a moment in answer and rose to tell the others, still feasting as the sun went down, of his decision.

Near midnight the *Zuen hildkoen-hitz* abandoned her body. Already she was in place, lying on her side, back against a stone slab embedded in the wall of the grave. Her legs were splayed to ease the pain of her twisted joint and her hands were clasped, for that was how they had often seen her speak to the unseen spirits.

The turtle whisperer arranged the shells over her body. It had taken him weeks to find them all: turtles were solitary creatures and hard to gather, and so they were precious to her. Now they had fed the people, and though they no longer had flesh they would give her their endurance, their patience, they would give her protection against evil from above.

The others threw in the remains of the oxtail and the wild boar, all that was left from their meal, to nourish her.

The girl placed the old woman's eagle wing by her arm. She placed the marten skulls close to the folded hands, and around them, as instructed, she placed the leopard pelvis, and all the rest along with a broken stone bowl, some bone tools, all that she would need.

Her task done, she turned away with relief. Now, freed, she could find a man. For the first time this day, amid the group's rising wails and moans, she smiled.

The others crowded around the grave. Some were crying softly; others tore at their faces or were solemn with grief.

Azkar could only watch, mute with pain, as they covered his *Em'kume* with earth.

Placed last in the grave beside her awkward leg was one final gift, the greatest of all, and the grave was sealed with stone. So that she might run in the other world, as she would need to do to continue to counsel and protect, she had asked for, and had received, a man's foot—a strong, healthy foot to give her speed.

Azkar would no longer be swift, not in this world. Instead, he would be the new *Zuen hildkoen-hitz* and the others would care for him, and protect him, and listen to his words, and follow his counsel.

A replica of the grave is on view at the Israel Museum. The Snail Creek Shaman was buried with all the objects in the story. Her skeleton had a severe congenital hip deformity, so she, like Drummer, must have walked with difficulty all her life. Beside her body was a human left foot. The story is there for the looking.

She lived at the beginning of that period during which human beings began to pivot to agriculture on a global scale. Climate had suddenly warmed, bringing in the Holocene, the current era. Plant life proliferated and the area teemed with animals.

Of course, the world was still dangerous and required management and explanation. The explanations, called myths, were the science of the day. They served humanity well.

With changing climate came increased population, which led to a new and lengthy dance among people, plants, and animals. The future would be very different. Once farming took hold, almost everything about human existence was different. High-starch grains became the foundation of diet. Domesticated animals evolved into dependents of human husbandry. Codified spiritual practices and an organized priesthood, who gradually replaced idiosyncratic shamans, exercised increasing political control. Monogamy and private property would eventually come to dominate a new social structure. That was still several millennia in the future, though. If the *Zuen hildkoen-hitz* saw the future, it was certainly not the one we inhabit.

Raising Stone

Around eleven thousand years ago the Natufian period was nearly over. Farming was taking root, so to speak. On a ridge overlooking a plain in what is now eastern Turkey people from many groups gathered, and for the next two thousand years invested enormous time and effort erecting structures in stone. They were circular like Natufian houses but much larger, with seating against the inside wall.

These were neither shelters nor houses. There is no evidence anyone lived nearby. Their purpose was entirely new.

They chiseled huge T-shaped pillars out of the local limestone with crude flint tools, hauled them to the top of the ridge, and raised them up. Over time, they built at least twenty such structures.

We know this place as Göbekli Tepe, or Potbelly Hill. Here people gathered in a time when they needed to find a way to smooth the increasing friction among them.

From his vantage point halfway up the hill Kemen gazed out over the plain. Even after ten days the clans were still arriving. Sun slanted over the tents dotting the green as far as his eye could see, casting long shadows toward the east. He grunted in satisfaction to see so many had answered the call. Never had the clans come so far to assemble like this.

Nekane loomed at his side, tall and severe. She stood in silence, drawing her long black pigtail through her hands. She dropped it and brushed her fingers across her high brow with a short laugh. "You are pleased, Kemen."

"Nekane! Look. You see," he pointed, "Gaïzka, our son, joins the stone-harvesters, young as he is." After a pause he continued, "As you requested, this year we build a Circle, call the spirits into us, all of us together. This year First Man and First Woman will reach the sky. They will soothe the tensions among the clans. This will be done for the Turning Moon."

Her strong nose flared. "How restless clans have become over the past twenty year-cycles!" she said. "They must bind together. The dark night of the Turning we will remake the world once more. They will help, all of them." She swept them all into her arms from the plain below, dotted with tents.

"Yes," he agreed. "Many will become People this season."

"And some will die."

"Some will die." He stroked his beard, recently streaked with white.

During the next two rounds of the moon, young men with flint tools chipped at the lines the carvers had drawn on the flat limestone of the hill. They painstakingly cut away several large T-shaped slabs a hand's length thick and an arm's length wide. Once they freed a slab from its bed, the young men levered it out with long poles. When it lay flat on the ground, the carvers felt along the surface for the outlines of boar or scorpion or water bird hiding in the stone. When found, they applied their flint tools to bring it up in low relief, creating an opening in the stone to set the animals free.

On a cool autumn morning of the new moon the elders came to consult with Kemen and Nekane. "These two," one asked, pointing at the largest columns. "They will live?"

Kemen tipped his head. "They are the couple that will dance face-to-face at the circle's center; their dance will lift the roof of the world."

The chosen band of young men arrived after a night of anticipation, when few slept. They set to work. First, they slid rollers under the two great figures and slowly pulled and pushed them to their final standing places.

The head carver, a short, wiry man in his early thirties, had already prepared a deep trench in the bedrock at the center of the circular wall. When the great T-shaped slab arrived, he snapped instructions at the young men. Because their languages were not always intelligible, there was a lot of talking back and forth, teasing, translating, explaining.

After some time they managed to align the base of the pillar with the slot in the bedrock. After a brief rest, and with much complaint, they pulled ropes and pried with levers. Slowly the great stone figure called First Man, tall as four mere humans, tilted up and thumped into place. By late afternoon, when the air was rapidly cooling, they had packed stones and dirt around the base, tied off the ropes, and propped the levers.

First Man, bound in place, loomed over them.

The carver thumped the pillar's side with a callused fist. "Tomorrow you will free First Man," he said.

They all know that once freed, the pillar would take life as a spectrally tall man, and his life would flow from the stone, from the carved parallel lines of his arms and elbows and the long-fingered hands touching the buckle on his fox-skin belt and loincloth on the narrow edge. The T crowning the pillar was a faceless head decorated with low reliefs of animals, intermediaries for the spirits between worlds.

"The next day First Woman will grow beside him," the carver murmured, gesturing at the stone on the ground many paces away, waiting her turn.

Nekane stood on a slight rise in the bedrock watching her son Gaïzka in front of the throng of youths packed close to the pillar, staring at First Man. Gaïzka's eyes glistened with pride, touched only a little with fear. This they had made, all the boys from the gathering clans working together. They had freed First Man from the stone, brought him to this place, and tomorrow he would live. Their labor had given such life.

Even as the carver spoke, a sound began, soft at first but quickly swelling to a deep bass groan. The boys shrank back against the surrounding crowd of women and small children. One of the women screamed. Others jostled in panic.

The sharp snap of a supporting pole stopped them. Splinters flew into the crowd and crimson stains appeared. The precariously balanced pillar tipped ponderously, gathering speed, and fell toward the crowd.

A fountain of gravel and dirt sprayed into the air from the base. First Man landed with a crash and split apart, sending up a cloud of pale dust.

The shocked silence continued. The cloud eddied and swirled and slowly settled.

Cries arose from the injured. One, a young man, lay beneath the broken column. He faced the setting sun, startled eyes wide open.

White foam mixed with blood covered his lips. One of the boys cried, "Gaïzka, son of Nekane and Kemen. It's Gaïzka!"

Nekane bent over her son. She could do nothing. The top half of the great limestone statue had crushed his body.

Beside her, Kemen showed her how the stone had fallen in such a way that the forearm on the broad side pointed straight at Gaïzka's head, while the hand hooked over the fox belt seemed to reach past it to touch the boy's face.

With a sorrowful nod, Nekane placed her thumbs on her son's eye sockets and began to chant his life back into the stone, to urge it to join the animals, the fox and the snake crawling along the broad face of the pillar's T-shaped cap.

Darkness gathered, pulling shadows into the slope of the ridge. The plain below was dotted with cooking fires. A sound swept down from the hilltop like a wind blowing through dead trees after a terrible fire. In this way the news of Gaïzka's death spread. Many faces would turn toward this place, but they were too distant for Nekane to see.

She called for bonfires. The older men would work halfway through the night levering the pillar halves onto logs and rolling them away so sculptors could reuse them for smaller pieces.

The initial shock of the accident faded, and others began to join Nekane's high, pure chant. One by one, in small groups, the people of all the clans joined, and their combined voices rose into a cloudless night sky. This great, spontaneous communal chant would ease the boy's spirit through the membrane of stone. The crescent moon crossed the heavens, and still they sang. Only at dawn did the voices slow to a stop.

Two men carried Gaïzka down the hill and out into the plain, and there by a small river they buried him amid lush grasses.

He was the first to die this season. He would not be the last. Always the people had to give back to the world that fed and clothed them. Always they had to attend to and measure the spirits' ways, know their meaning, how they helped or hindered, how they gave health or sickness, how they distributed pain. Gaïzka's death was rich in meanings: First Man, the great being of stone, had killed him, and in doing so, First Man had died as well.

Gaïzka had given his life to the stone before it was fully alive. Nekane and Kemen talked over this puzzle most of the day while the sun passed overhead.

All who had made the fallen First Man must make another even taller than the first to complete the joining of the clans. Only in this way could they create harmony.

To that end Nekane and Kemen discussed and carefully selected what the new First Man must look like, and how it would bring people and spirits together through the animals he wore.

So it happened several days later that the inert stone form of a new First Man arrived at the ridge top. His body was decorated with square-jawed lizards, a row of ducks, and a bristle-backed boar with lowered head. A scorpion scuttled along the one side and a snake writhed up the other, while a fox with a prominent erect penis stared out from the top just beneath the great T of his head. On the head itself a single vulture spread its wings.

Nekane and Kemen sighed with relief when the column settled firmly into place.

At dusk they walked down the hill to the others, to the young men, the carvers in stone, the shamans of all the clans, and all the people.

"Vulture will watch over our son," Nekane said.

"As it must be," Kemen murmured.

At the bottom of the ridge, where the tents began, Kemen stopped. The day was still and cool, with a pale sun. The people looked at him expectantly.

"Before this season ends and the clans disperse," Kemen told them, "We must bury First Man, First Woman, the circle. All must return to earth."

The word flowed back from those in front to those in back, for the people assembled knew this was, as always, their final task until the next gathering.

Most of the large, fearsome animals of the Ice Age, the mammoths, cave lions, and rhinoceros, were now scarce or extinct. The low-relief carved animals at Göbekli Tepe are small—lizards, ducks, wild boar, fox—marking a shift in the relationship between men and animals. Here the animals are on a more intimate, human scale. Many, like scorpions and snakes, were still dangerous. Others, like ducks and lizards, were not, but people had a highly emotional engagement with them, whether fear or awe. And they were still wild. Humans had not yet gentled sheep and goats into placid domesticates unfit to survive on their own.

Rabbits and water birds provided food, but the spiky, poisonous ones, or the carrion eaters had other meanings, more personal, seldom seen: spiritual. They emerge from, or pass through, the stone with less of the terrifying power or life energy of the cave paintings.

They may have represented totems of the nomads or newly sedentary communities in the upper Euphrates Valley now forced to negotiate competition and cooperation.

Göbekli Tepe was a large-scale communal project, the beginnings of a new kind of separation from the environment, a temporary built world.

When they buried the structures at Göbekli Tepe for the last time, the lure of farming was growing strong. Hunting and foraging had not completely disappeared, but their importance had begun a long decline.

Ritual—the building of the circles; the songs and stories of origins, hunts, or great deeds chanted or recited inside the circles protected by enormous columns; the flickering firelight on the spectral couple of spirit-stones that dominated the circle; the labor of young men and the awe they felt before those parental statues—would help bind the clans.

Göbekli Tepe represents a major spiritual shift and hints at the deliberate invention of religion, not just its emergence. Places set aside for special ceremonies were necessary for agriculture to take hold.

Hunting grounds had to be shared with more strangers. Ceremonies could seal agreements about the use of the land. Something called "ownership" was emerging. Nonhuman things were growing apart from human ones. Göbekli Tepe may have created a community of communities that could, for a time, transcend a growing animosity among competing strangers. The long ages of mankind's existence in the world, though sometimes harsh, were innocent, intimate, and interconnected, in their way, a paradise. With a commitment to agriculture came separation from the natural world. Mankind was not *condemned* to a life of perpetual toil, this was the price of becoming the Lords of Creation.

Feasting

After a 1,300-year cold snap called the Younger Dryas, 12,800 to 11,500 years ago, warm weather returned for good. Settlements expanded into villages and the need to form and nurture broader alliances with shared food and celebration grew more urgent.

To this end special places designed to unite local and regional farming communities emerged. One such place is Kfar HaHoresh in Israel, less than half a day's walk from a number of settlements.

Lond, a crone of more than forty years, waited with the other adults. The boys were still out of sight, but their shouts were growing louder as they moved up the wadi toward the cattle trap under the ridge.

The panicked animals thundered into view, followed closely by her grandson Bemman and the others, waving their arms and shouting.

With escape cut off, the panicked cattle turned back. Some scattered up the sides of the wadi and escaped into the scrub, leaping and staggering in comic desperation.

Bemman was bent over laughing at them. Always the leader, he threw stones at their bony backsides, one after the other, sharp-edged flints and baked mud balls. He turned to wave at his grandmother when a bull stopped, ready to charge the young men. Bemman ran in and twisted its tail. When it swung its great horns around, he jumped away, capering around. Dust hung thick in the warm air.

For the families gathering from as much as a day's walk away, this was a happy time of chaos and violence, a welcome change from the drudgery of farming, hunting small animals, gathering fruits and tubers, not to mention constantly repairing their fragile huts. Only these great feasts broke the monotony of settled life.

Only now that the harvest was in could they feast as they deserved. Through the long summer Lond had watched Bemman. She had time, for she was old and dozed in the sun. On slow evenings she watched as he and his friends fashioned their clumsy spear points and practiced the dodging and dancing moves they would need.

As the cattle tired, the stronger boys threw their spears. One enormous bull turned on them, they scattered, then ran back in again. Bemman came close for a quick spear thrust.

This time, though, he was too slow. The bull's massive hoof caught him in the side and sent him tumbling. His head struck a large limestone block left over from last season's quarrying for the platform on the slope above. As soon as the last beasts had rumbled on, three or four young men carried him up the west side of the wadi to the glade. The others attacked the bull with renewed energy and brought it down, bellowing in pain.

Lond squatted by body on the ground beside the broad platform. Its plaster glistened, still fresh and damp. She leaned down and looked into his vacant eyes. "Grandson, our boy," she muttered. "You came too close this time. Poor boy."

The boy's kin gathered around the broken body. Lond waved a crow's feather over his face and shattered skull before launching into a low, rasping singsong. The family stared without expression.

Those not in the immediate family backed away. At first, they watched from a distance. The boy would never move again, though, and they returned to work preparing for the feast. Such things happened; children died too often. Lond waited, grieving, then left him. There was too much to be done.

The boys dragged the slaughtered cattle to the eastern side of the glade where the cooking pits were already burning. The bull, several cows, and a calf were laid out for butchering. Men and women, including Lond, worked side by side to strip the skins, cut meat away from the bones, and separate the larger joints for roasting.

While the meat sizzled, Lond and some of the younger women pounded the large bones with rocks, cracking them open to extract the marrow.

She seemed to be listening to the young mothers' gossiping of liaisons, runaways, and deaths in their villages to the north near the inland sea. She seemed to have forgotten the boy, already in the past. But she had not forgotten.

Next to the marrow-mining, the older men gathered under the trees were working stone into blades, scrapers, and other tools. Their hammerstones struck a range of ringing tones from flint cores and filled the glade with rhythmic music. The sun dropped behind the western slope, and the glade blazed with firelight. Smoke from the roasting pits overwhelmed the thick odor of coagulating blood.

Others from more distant settlements were arriving in a steady stream. They crossed the blood-soaked ground under the limestone escarpment and joined the feast.

The moon flooded the glade with silver. Drums, whistles and rattles joined together, weaving a new, ever-shifting sound tapestry that echoed the earlier rhythms of the stone-working. Through it all Lond sat a little apart, thoughtful. Her only grandson was no more, and there was little hope for another. What did this mean to her family? She pondered the question, and was not sure.

The feast continued late into the night, and still she sat apart. Once the final bones had been discarded and the children scattered around the platform could no longer stay awake, the adults took up the music and dancing again. This time it was bawdier, with outbursts of laughter, teasing shouts. The darker patches far from the fires were for serious courtship. The music, the pounding of drums, plucking of strings, and shrieking flutes grew wilder and more disorganized. At dawn the music died away and some of the people with longer walks ahead of them left.

Only then did Lond stand, dizzy for a moment before she found her balance. The rest of her family was waiting at the far end, seated around Bemman. Although he had died in service of the ceremony, and would be honored for that, he was only one of the dead to consider. People from other villages had brought the bodies of their kin to join their ancestors in private graves or a place of honor beneath the platform. They spoke quietly together until all the others had left. Burial was a private thing.

The dead boy's family looked to Lond. She had them scratch out a shallow space in the plaster surface. The ground beneath was hard, and they soon uncovered the skeleton of a girl who had died some years before. In death she had worn a necklace of shells. "I remember her," Lond said. "Her name was Vezel. She had eleven years."

They moved her to one side and put Bemman's body in her place. The ground was hard, so the grave was small. Lond and the boy's father flexed him for burial. The day was beginning, there was work in the settlement, and there was not much time.

Bemman's father fetched from the wadi the limestone block that had killed him. He placed it in the grave and stepped back. Everyone in the family touched the body as they laid his head on the stone. Lond fingered her own necklace of shells. She felt something was missing, something important, but could not think what it was.

They covered him with dirt while Lond mixed lime and water to plaster the grave. The lime glowed white and clean. Next spring they would still know where Bemman was buried, but its location would fade, as had the girl's before him. Only Lond would remember.

While the plaster dried, several of Bemman's kin spoke of his courage, his laughter, his brief life.

When they finished speaking, and all except those with dead to bury were about to leave, an old man seated on the west side of the platform said, "One more thing."

They turned. With his woman's help he pushed himself to his feet. "This is the year," he said in a quavering voice. His words were muffled, for he was born in a distant place. "I must return my brother's skull to this place. It has lived in our house for a year now, and it is time for it to rejoin his body."

His woman handed him the skull. The old man bent down and placed it at the base of the western retaining wall. He stood up with a smile. "So it was done in my home when I was young."

Lond had watched the old man take his brother's skull, lovingly plastered over with squinted eyes and pursed mouth, and place it at the corner of the platform. She knew then what she had missed all this day.

Next year, she would remember Bemman's place under the platform, where he could not see what happened above him. Next year she would retrieve his skull. It would be clean, ready for her to do what this man had done with his brother.

She would give it a new face, and after another year watching over her family's home, if she still lived, she would bring Bemman back to this place. If she did not live, she would pass that task to someone else.

She, or they, would place Bemman's skull so that it, too, could look out over the platform and enjoy the feasting.

Ten thousand years ago settlements along waterways in lower Galilee were small collections of houses with plastered floors. The inhabitants farmed, foraged, and hunted. As was widely practiced in the region, villagers buried at least some of their dead under their floors. They made body ornaments and farm tools, figurines and lime plaster. As farming took over days and nights, opportunities to meet outsiders became more and more limited.

The north-facing slope at Kfar HaHoresh was too steep and rocky for farming but high enough to attract large groups for social interaction, sheltered yet accessible. The opposite hill offers a broad panorama of the lower Galilee, but this glade was enclosed and private. Over a couple of thousand years people constructed a complex series of plastered platforms and retaining walls. Sometimes they framed the platforms with one or two courses of stone. They buried some of their dead here and set stones to mark graves. Later they removed and reburied their skulls. They left the remains of kilns for making lime plaster, pits for roasting meat, collections of animal bone. There were posts to support shade or cult images.

As networks of relationships grew more complex, so did disposal of the dead. At Jericho to the south, families dug up their skulls to model in plaster and sometimes paint elaborate new faces. As the dead grew more demanding, ritual around them enforced more durable memories. Death was woven into daily life.

The human species may not have fully committed to farming yet, but our own domestication was well under way.

Waters Bitter and Sweet

Humans, ever more numerous, were spreading over the globe. Some ventured onto the seas and out of sight of land. This happened in Indonesia and Australia sixty thousand years ago.

It happened in Europe, too. As much as twelve thousand years ago people crossed the bitter salt water to the island of Cyprus, a place unusual enough to generate its own kinds of stories.

As vapor, water swirls overhead only to thunder down as rain or gently shroud the land in mist. As a solid, it tempers the earth like a metal that resists opening, shifting, or exploiting. It once towered over all who ventured close, blocking their way, driving them before it, always southward.

As liquid, water gurgles and flows, roars and simpers; it takes away life as readily as it gives it.

People gathered on the cobbles and sand to gaze longingly across tumultuous, bitter, undrinkable water at barely visible, seemingly unattainable lands. Over time these distant islands grew ever more tempting. Though the sea before them contained treasures of shellfish from the rocks and gray triggerfish that could be pulled up with a line from the depths, the people wondered what lay on the other side of the water.

One day a group of them set out on a raft with water, clothing, animals, and stone for tools. By trial and error they learned to follow the seasonal changes in winds and currents until they reached a great island. When they turned to look back, they could no longer see the land they had left. Years passed, and the journey became routine. They survived and flourished.

After many years, after many centuries, a man stood on a platform on the side of a slope. He was ancient, or so his audience believed, and looking at him you would have to agree. After all, thin hair feathered out over a high forehead, and his cheeks were sunken.

Old as he was, though, his eyes burned with a wicked humor that terrified the young people seated around the platform. To cover their fear, they joked among themselves.

The sun, halfway down the sky to his right, bathed his cheek in honey light, seeming to ignite each wispy hair into a tiny torch.

On a cleared slope behind him the adults lounged on other irregular platforms of stone and cobble, plastered over with the white dust that grew in the earth. They were knapping chert or flint and gossiping in low voices.

In one hand the old man held a long shard of transparent stone. He held this up so it caught the light, and he waved it back and forth. His audience stopped talking and looked at him.

A light breeze had set the old man's hair to gently fluttering. He brushed a hand over it, but it sprang up again. "This stone came with the first men," he said.

They could see the sun burnt gray-green through it.

"This stone from the guts of a fire mountain was given me by my teacher. It is why you listen, is it not, because I hold this?"

They nodded; eyes fixed on him.

"Very well." After a moment's pause, he began in a new tone, "You will remember we killed the last cow." His lips turned down in a rueful expression, for it was he himself who had slaughtered her. "That was soon after the earth shook and Anthar's house fell, four cycles ago. She was the last cow, the last we have seen here, and she provided a meal to remember. We will never have another such; these lands are not kind to cattle. You can see down there how dense the woods are, the pistachios, the almonds, the pines. There isn't enough for cattle to forage, and we barely grow enough food for ourselves. My father's father brought cows on a hard crossing, two days of rough passage beaten by wind, tossed about by waves. Cattle survived the journey, it is true, but they did not thrive. And that last one, she was half wild, and had killed a boy. You remember that, I'm sure."

The children playing in the meadow below had looked up when the old man started speaking, but soon lost interest and bent back to their games—old men were always talking! The young men and women around the platform, though, they nodded. They were old enough, and remembered the earth shaking, the falling stones, the deep, terrifying rumble. It was the anger of the earth, people said. Like the sky, the earth had its fits; anger burned in it as it did in man. The earth had her own desires, her own needs. Sometimes men disturbed her. People whispered then they never should have brought the cattle. The land did not like cattle.

The old man swept his hand, taking in the bowl of scrubland to the south, the circle of hills opposite, the glint of river. "This land gives us much," he went on. "We live beside the waters. We hunt birds at the Salt Lake. In summer we come into these hills to hunt the deer and pig as those who came before us have always done. We have our sheep, our goats. Some of us grow wheat, and we make our bread, our delicious soup. Yet, this land is not always kind to us. It shakes with anger. It is sometimes cold, sometimes hot. But even without the cattle it is mostly kind. But it was not always so. Things were different in ancient days."

He stopped. He seemed to have finished, but his listeners knew he had not. It was a trick he had. One of the girls said, "Tell us. What are you waiting for?"

The old man nodded, lips pursed, as if thinking. "I wait for the sun." They laughed then, for the sun went its own way whether men urged it to hurry or begged it to linger. There was no point in waiting for it.

"Very well," he agreed, straightening his shoulders. "The story I tell you now is of times before men, for I think you do not know the true story of the giant."

His listeners leaned closer, drawn by the way his voice fell into a special rhythm, the way he paused before announcing his subject. No, they did not know the true story of the giant, only what their mothers had threatened when they were little. "The giant will come in the night," they would say. "He will eat you. You can still see the bones of naughty children if you know where to look. One day we will show you, if you survive long enough."

Remembering these dire warnings, they all felt the thrill of fear. Their lips parted a little and they licked them. "Tell us," the girl said.

The old man grunted with pleasure. "You are Tyllen."

"Yes, Elder."

This would be the girl he would teach the stories, to remember and tell. He would watch her closely another year, this scrawny child of twelve with bright eyes that saw everything, but in truth the teaching had already begun.

"It was in the age of mists," the old man began. "Before the time of men, as I said, before the days of toil, when the world was as it always was. Not like now. Not like now at all."

He watched the girl mouth the words after him, *Before the time of men... Not like now at all,* and a smile passed over his lips. But he frowned sternly and continued. "In those days a giant's stride carried him across a river in one step. His arm reached around hills. His breath summoned storms, blew almonds from the trees, tore furrows in the earth."

He had them, all eleven of them, leaning forward, chins in cupped hands, elbows on knees. "One day, when the mists rose highest and cold crept up and turned the mountains white as the plaster on our platforms, as the dust of the earth, the giant decided to leave where he was, for in the distance he saw land no one had ever visited."

"This land," Tyllen murmured.

"Yes, this land. We were not here then, of course."

"Why did the giant come?"

"Why? Why do you think?"

"He was hungry?"

"Well, he was a giant, and giants were always hungry." The old man pulled in his lower lip, the way he did when he was thinking which way to go in a story.

He lifted the obsidian core and began again. "So, he built a boat, though he could have walked, of course, being a giant. What he built was not a boat like ours, not a flat shelf that made men sweat over oars, but an enormous hollowed cedar log that could swim across the bitter salt like a swan. So swiftly did he come that instead of taking two days and a night, as we must, he was here even before the sun reached the overhead. This was in the days before men, as I have said. And do you know what he found?"

The young men and women shook their heads, fingertips in their mouths, fishing for words on their tongues, for answers, all except Tyllen, who only waited.

The old man thought over the next words. He could say this, or he could say that. "No men," he told them at last. "There were creatures here, but they were not men, they were monsters, so I learned from our storyteller, and she from the one before her. Monsters larger than men lived in this land. The giant saw them, and though the monsters were smaller than the giant, he was still afraid, for these monsters had mouths that opened and kept opening, and could swallow even a giant in one gulp. Their harsh, guttural voices could topple trees. They could take a grip with their wide flat teeth and pull an oak from the ground, so it was said. And when they saw the giant gliding up onto the sand, when they saw the giant step from his swan-boat onto the beach, they swarmed in their hundreds with their white teeth, their dark, piercing eyes, and their glistening black skin. They made a river of darkness flooding swiftly down the cliffs facing the sea. They flowed onto the beach where the giant had landed near the Salt Lake. They did not stop there, no. They came on until they reached the water's edge. They tore the giant's swift swan-boat apart and devoured the shreds. Oh, they were fierce."

He stopped again.

"What happened then?" Tyllen asked, and others repeated the question.

The old man leaned forward, jutting his tufted chin at them. "Then? What do you *think* happened then?"

Before anyone could answer there was a roar of greeting from the people above, and they all stood up, shading their eyes. Hunters were coming down from the trees at the top of the slope carrying a deer slung on a pole. The old man sighed; his stories were so often interrupted by food. Well, so it is.

Even the smaller children dashed up the gravel pathway winding among the platforms. They ran through the tall grasses, brushing the tops with their hands, and gathered around the hunters. With the three camp dogs barking excitedly at their heels, and squealing with delight themselves, they capered around the hunters all the way back to the camp. The hunt had been bad for some days and they all craved meat.

The children shrieked with delight as two of the women cut the deerskin with sharpened stones and pulled out the guts. The children leaned over the women as they rolled the hide away from the meat.

A fire blazed beyond the edge of the camp, and the whole band prepared for a feast on the spot, tools dropped in place, forgotten, and the story of the giant, also forgotten. Someone fetched sacks of water from the creek and began heating it for grain gruel. Oh, it would be a great feast. Even if they no longer had a cow to kill in its honor, still, a wild stag would do. Already one of the hunters was cutting away the antlers.

The women sat in a circle, scraping the inside of the hide to remove fat and tendon. As they worked, they threw bits of fat at each other, or wiped blood on each other's tunics, whooping and shaking with laughter.

The men took joints and slabs of meat to the fire, and while it roasted three others step-hopped onto one of the platforms and began a slow circle walk, heads back, looking side to side, and everyone recognized the great strutting bird when it gathered in noisy groups in late winter. More men joined the dancers, displaying their courtship plumage. Even the old storyteller could see, dim as his eyesight had become, the way the dancers held their hands behind their backs and waggled their fingers like a tail sticking up, the way they sank their heads back into their necks, the way they stepped, one foot forward and down, then the other.

A few of the unattached women waited just off the platform, backs to the dancers, their own finger-tails up and waggling. One of the men hopped off and a woman ran away up the slope whispering through the tall dry grass. The man ran after her. After two circuits of the field they collapsed into laughter while the others applauded, hooting and grunting in imitation of the bird. "If we only had one for supper," someone said pointedly. One of the dancers came over and offered his neck to her, and as the group dissolved in laughter someone else said they had a fine deer to eat, and deer was even better than a bird, no matter how large.

The day had begun to cool when they approached the fire, sniffing at the scents of roasted venison. One man cut hunks of meat and the others, seated now, passed them around. One of the cooks cracked open the long bones, dug out the marrow, and put it into a large stone bowl. Men and women fed each other, or sniffed at each other in raucous play, or told sly, erotic jokes with their mouths full. They scooped out fingers full of marrow, and grease glistened on their faces.

As the festivities settled down someone improvised a melody, stringing nonsense words together at first, and then listing the hunters' names, extolling one's skill at throwing the bolo that tangled the deer's feet, and then the strength of the men who bludgeoned it to death, and the ingenuity of those who tied it to the pole and carried it back.

Someone else picked up the thread and sang of the fire, which had come to life out of the wood and was giving of itself to feed the people. Then the song turned to a hymn of praise for the fire, and then for the deer, and for the earth that nourished the deer, and for the sky that arched over them all, and for the happy meeting of the men with the animal.

Someone repeated the names of the hunters in a higher register, and others joined in, clapping hands, and the names became a refrain, and the words become more erotic, more playful, and gradually the song faded away and silence fell over the group.

The couple that had circled the field was now face to face. The silent group watched as the man took an almond from his pouch and held it out. She leaned down with her mouth and took it in with a slurp, and everyone smiled, for in this way she had accepted him, and they left together.

After that the rest of the adults began to drift away into the darkness, some with a partner, others accompanying small children, yet others alone.

Only the old storyteller remained by the fire, staring into the flames.

Tyllen approached. "Would you continue, elder?"

The old man arched his eyebrows in mock surprise. "What? Finish? For you alone, you mean? Or for you and someone else, girl? I'm a busy man. Come back tomorrow."

Tyllen smiled. "Don't worry, we will all be here. Just say yes."

"Hmph. Now you mention it, the fire will do nicely, for light, of course, so I can see faces. Stories need ears to hear, you understand? And eyes to watch. Listeners tell the story as much as the storyteller, who learns from their faces as he tells."

Well, he thought, the teaching might as well continue. The girl was eager.

Tyllen quickly disappeared into the darkness, leaving the old man to dream in the flames, but almost as quickly as she had left, though, the girl returned with the others.

"So soon?" the old man asked, and for a moment they believed he regretted their coming.

He rose painfully to his feet. His listeners, as he called them, made room for him to stand where he could see them all. The fire was dying down, and the red half-light flickered on their faces turned up to him, waiting.

He took a stance, feet apart, and crossed his arms. "Mmm, hmm," he murmured, as though fumbling in his pack for something he had lost. "Where was I?"

"The monsters ate the giant's swan-boat," someone said, and the old man saw the look of approval on Tyllen's face.

"Yes, of course," the old man agreed. "I asked you what you thought happened next; at least, that's how I remember it, hmm?"

"Yes, yes, that's what you said," they responded.

The old man stroked his chin, tugging at the few hairs of his beard. "Well?"

"You really want to know what we *think*?" a boy asked. The old man saw Tyllen smiling. The girl understood what this was about, this teasing.

"Of course," the old man snapped. "That's why I asked."

"Well, the giant lost his boat," a girl ventured. "How did he get home without a boat?" She looked around at the others. "Well, there are no giants here now," she explained.

Another boy suggested the monsters ate the giant. "There were a lot of monsters," he said. "Even though they were smaller, there were a lot of them, and they had huge mouths. They could have eaten the giant. You said so."

The old man shook his head sorrowfully. "No, they did not eat the giant. But you say one thing truly. The monsters did turn on him. The giant, with no way of returning home, was angry and there was a terrible struggle. It was like the stags butting heads together when they fight over a doe. It was such a fight, except unlike the deer, there was real death, real pain. The mating fight of the deer was nothing to this, the charging, the backing away, the threats. The giant ran, and stomped, and roared, and the monsters squealed their rage, they swarmed at the giant, and pulled him down, but he rose again and shook them off, bellowing. A day passed like this, in blood and rage, and another. Seven whole days went by before the end. It was like a hunt in that way, that the end was hidden. Clouds of dust darkened the sky. Screams of pain shivered the stars at night and made the sun shake in the day. First the struggle went one way, and it seemed the giant would crush the monsters. The next moment the monsters would swarm over him again and all would change."

He stopped with a benign smile at the circle of faces in the flickering firelight. Then he slid the obsidian core into his leather pouch and sat down.

"You can't do that," one of the boys complained.

"Do what?" he asked innocently.

"Not finish."

The old storyteller pulled his head back in surprise. "Not finish? But you know the finish. There is no need to tell it."

"Who won?"

Tyllen said quietly, "It was a great struggle. In the end the monsters were killed by the giant, and the giant vanished from this earth."

"But ..."

"Listen to Tyllen," the old man said. "Your mothers threatened to show you the bones of naughty children, killed by the giant. What they would show you are the bones of the monsters." He squeezed his eyes shut. "Or, perhaps, of giants, if others came looking for him and perished here," he added. "Who can say?"

While the Natufians were building round houses in small hamlets, people traveled to Cyprus, the third largest island in the Mediterranean. What they found was a wondrous paradise of wooded slopes, rivers, mountains, lakes both salt and sweet, almond trees, large, ungainly birds, and strange pygmy elephants and hippos.

On the map Cyprus resembles a flattened soup ladle, its handle extending toward the corner where Turkey and Syria meet. Today it is 65 kilometers south of Turkey and 105 kilometers west of Syria, a wrinkled layer cake of old seabed over volcanic rock. It may be a weathered, seamy, pale, striated land, but it is still well watered and favored by climate.

People rode down the currents from Anatolia or the Levant, bringing with them their domesticated sheep and goats, and even some cattle. They brought wild pigs and Mesopotamian fallow deer. They grew peas and lentils and emmer wheat, gathered native almonds and pistachios, tended their animals. They hunted their imported deer and wild boar in the foothills of the Troödos Massif. They built villages of round houses with stone foundations.

There were no hearths or house foundations at Ais Giorkis, just a seasonal camp of stone platforms with processed cattle bones and plants both wild and domestic. And burials. It was a place for stories.

In the old man's tale dwarf hippos, called pygmos, were the monsters. With no natural predators they were unafraid, and though they weighed two or three hundred kilos, they were easy prey to the first wave of humans.

A rock shelter above the sea on the south coast contained the bones of five hundred pygmos mixed with stone tools. People, driven by curiosity, perhaps, or by the clamor of their increasingly sedentary world, had come across those strange, fleshy, slow moving animals that offered easy protein. They processed the animals and without a doubt facilitated their extinction.

Those first visitors also vanished. Like the horse hunters at Solutré, they were profligate with resources and became victims of their own success. Those who came later could only fashion stories of giants and epic battles to explain heaps of extinct animal bones.

II

HOUSE

Frozen Tears

Ten millennia ago the progress of farming in the Near East was reaching a point of no return called the Pre-Pottery Neolithic B. A prominent feature of the PPNB was the square house. From this moment on nearly all settlements would build exclusively square.

The ad hoc, long-range trading of the Paleolithic, when a hunter-gatherer like Nyla could have a shell passed from one band to another in a random journey through Europe, was gone. Social networks had expanded beyond local feasting and isolated settlements. Necessary resources like obsidian moved over well-traveled routes. Humans were organizing themselves, their bodies, and their landscapes in new, more intricate ways.

The woman approaching from the west, bent under the weight of a basket strapped to her back, believed her settlement in the curve of the river was the center of a world that extended in all directions, even beyond the distant mountains.

It was late summer, the day was warm, and crossing the river she smiled at the tug of it around her ankles and calves. The current had slackened during the dry season, and the cool water refreshed her. When she reached the bank, she paused, reluctant to step onto the dust of the street. To give herself another moment she wiped the back of her hand across her brow and squinted at the houses.

She was hoping she might see Lokke coming back from the mountain to the east. He had been gone for three days and it was almost sundown. She imagined him walking down the dusty path with a satchel filled with the mountain stone her people called frozen tears, for he was foremost among the stone hunters. So great was his skill at finding the precious nodules from which they made all their tools that she was sure one day his descendants would be called Frozen Tear clan.

She knew how his face would brighten when he saw her. He would call her name, "Semmi," in his gentle voice. "You look beautiful today." He had repeated those words so often she had allowed herself to believe him, and even though today his clan was small and his hunting indifferent, he was a good provider because he went often to the mountain to bring back nodules of the shiny black stone. Everyone liked him, too.

He could have been there and come back by now, she thought, but he was nowhere in sight.

She shrugged her tall basket of wild grains against her back. She had spent the day collecting them and was happily tired, walking along the road between the houses and the common buildings. Her house was near the east side away from the river, for her family had been among the last to build. While the house was some distance from the river, it was close to a productive stand of hackberry trees.

She passed rows of adobe brick drying in the sun along the riverbank. The brick smelled sharply of baking clay, late summer, and home. Everything this time of day, this time of year, made her smile, for she was still young. She waved gaily at three women making wool yarn on their roof by the road.

Some distance past the common buildings her smile vanished at a woman's piercing cry of pain, distorted with horror and fear and wild protest. She dodged quickly into an alley between the houses. Her basket knocked against the walls, splashing grain. The cries continued, ever louder as she approached.

She pushed her way through a thick crowd into a small courtyard. "What's happening?" she demanded of an old woman, one of Lokke's maternal aunts.

"Lokke's sister." The old woman waved a thickly veined hand to ward off misfortune. "Spirit Man is in there." The screaming continued, broken only by the sucking whistle of indrawn breath.

"Her mother's been saying a spirit moved into the girl, it's true," Semmi said. "Especially since her baby died."

"A bad spirit," the old woman agreed. "That's why her man wouldn't stay, whoever he is. That's what they say. The poor girl scratches her face; she hits her head in grief and screams if anyone tries to take her baby. So her mother asked for Spirit Man."

Semmi climbed to the roof and looked down into the house. The fire for dinner should have been going in the hearth, but she saw only cold ash. She could see Spirit Man's back bent over something.

Lokke's mother moved into the light from the opening. "Semmi!" she called over the high-pitched screams. "Come down, please." She was visibly trembling.

Semmi shrugged her basket to one side on the roof and climbed down into the house.

It was dark and stank so much of fear and sweat and blood that for a moment she could see nothing but Spirit Man's back and Lokke's mother's trembling lips, but soon she realized that Spirit Man had the girl's head wedged face down between his knees. He had cut open the skin of her scalp and turned the two flaps back. Blood flowed freely, soaking the girl's dark hair.

At his side was a small leather pad with a collection of clear black cutting blades. He was using one of these to scrape a spot on the back of her skull. Already he had made a small hole in the bone, and despite the gloom of the interior and the flow of blood Semmi caught glimpses of pink brain. She had seen enough slaughtered animals to know.

He kept enlarging the hole, cutting away small flakes of bone. When it was the width of a thumb, Spirit Man put aside his instrument and leaned back.

The cries stopped. The girl had fainted at last.

"That's enough," he muttered, mopping up the blood with a scrap of wool. "Bad spirit should be gone." He deftly folded the flaps of skin together and wrapped a leather band around the girl's head to hold it in place. "You should bury that thing," he said, nodding toward the infant corpse in the corner. "Now, let her sleep." He pulled himself painfully to his feet and brushed his leather apron. "When she wakes give her a broth of willow bark, nothing else for three days. She may live. She may not."

He rolled up his instruments, climbed the ladder, and disappeared.

Almost immediately the light dimmed. It was Lokke. Semmi stared into his eyes but he didn't even notice her. "Why did you call Spirit Man?" he demanded of his mother.

"I couldn't let her suffer any longer," she said.

He made a spitting gesture. "You mean you couldn't let yourself suffer any longer," he snarled, looking down at the sleeping form.

"She's been getting worse."

"It's true," Semmi put in. "She was scratching her face; you can see where she did it. Saying wild things that made no sense. Trying to hit people. You know all this, Lokke. Especially since the baby … ." She tried to make it sound gentle, but she heard resentment in her words. Was it because he wasn't there to greet her when she returned from the stands of wild wheat down the other side of the river, or because he hadn't looked at her just now? Was she already thinking of him as a husband?

He ignored even these words, and that hurt worse than his earlier absence. He had turned away from them both. Now he dropped his chin and shoulders and sat down beside the unconscious girl. "I know," he murmured, taking his sister's hand. "I know she was suffering, but … ."

Semmi heard his sob and turned away. She could see the color fading from the girl's face, could see the life force leaving even though the malicious spirit had left through the hole in her skull.

When they buried the dead girl with her infant six days later, Lokke's mother asked Semmi to assist. She treated Semmi as if she were already joined to her son, though Lokke had said nothing of it, and even seemed indifferent. She managed to overcome the strange new fear and panic she felt in his presence. Later she helped wash the body, feeling it grow stiff as she bathed the girl's feet. She, Semmi, was the one who blew the ochre into the girl's mouth, though she was no longer confident the red color would really help the dead girl continue to live in the other world. It was she who helped Lokke's mother clean out the house, scrub the floor, and dig into it to prepare the grave for mother and child. And it was she who said the final goodbye and helped cover them. Together she and Lokke's mother smoothed new mud on the floor.

Lokke did little to help. He had dropped his load of obsidian outside the house, where it lay abandoned for days. Little by little people came in the night and bore it away until there was none left. The day after the burial he left with some of the men. Four days later they returned carrying the butchered meat of an enormous aurochs. This time, as they announced a village feast in the building across the main street, Lokke looked at Semmi for the first time; it was the look of a successful hunter.

No one mentioned the dead girl at the feast. She had gone under the floor of the house, a private person who no longer belonged to the village, but to her family only. More and more, as Lokke drank fermented grains he looked at Semmi, and spoke to her at last, if only with his eyes.

That is when she began smiling again as well.

Obsidian makes an edge sharper than any modern scalpel. Ancient volcanoes are a rich source, and Lokke's town was central to its trade. For five thousand years, long after the town was gone, the local obsidian traveled as far as Syria, Palestine, and Cyprus.

Lokke's people experimented with tending barley and wheat, but Semmi still collected wild grains at the river where they grow to this day.

The floors and some of the walls of the ceremonial buildings, and they were now buildings, were painted red, drenching social spaces in symbolic meaning and memories of past events.

A skull found on site is the earliest known example of trepanation, the drilling or scraping of a hole in the cranium either to relieve pressure, as a cure for chronic headache, or to release evil spirits, mental or emotional disorders. Such physical tampering expresses a shift in the human relationship with the non-ordinary world.

There are few of the tiny crumbs of daily life normally found in such sites: no grain in the storage rooms, no paintings or fixtures on the walls, no decoration, and a paucity of made things like figurines. The stone and obsidian tools were all imports.

Burials under the floors are proof of occupation, though. When the inhabitants moved away they took almost everything, leaving only their dead to connect us with their past.

Curves, fractals, layers, and cycles had long dominated vision and consciousness; now square houses made of sun-dried brick began the long trajectory of the straight line. Over time, as houses clustered ever closer together, they pushed their circles out into conjoining squares. Lines, easy to make from sun-dried brick, would eventually separate people from their leafy, visually complex, vibrant natural home. Inside would be us, outside would be others. This squaring of the circle changed perceptions of the world. Even today the straight line dominates everything. Just look around.

Nearly everyone in the world was busy building permanent, multigeneration houses. Their very existence flattened the cyclic model of time into a straight line from past into future.

Tracks

Unpredictable events—solar eclipses, devastating earthquakes, comets, terrible storms, epidemics, or volcanic eruptions—had always haunted the human world.

Such events were conduits for vast, incomprehensible forces inimical to man. What other explanation could there be for such unprovoked violence than invoking the enmity of the supernatural?

These forces were not yet gods, but already they had the power to announce a death, or cause it. This is the story of a death announced.

Once long ago (so the bard would sing), a light dusting of snow covered the houses during the night. By morning, though, the clouds had fled, leaving behind a day shimmering with crisp blue clarity.

The sun was full above the eastern hills when Semmi, her gray hair tied against the nape of her neck, made her way down the broad gravel road separating the clumps of houses on her right from the common buildings to her left. This was the shortest day of the year, feast day. Smoke from the large community ovens rose straight into the sky where it thinned away to nothing. Already hints of boiling semolina floated on the air.

Beside her skipped her grandson, the one who asked too many questions. "When will grandfather come back?" he asked, and without waiting for an answer said, "Can I go with him next time?" And again without waiting asked, "Where are we going? Will he bring frozen tears back to us? We're the Frozen Tear People, aren't we? Aren't we?"

Before she had children Semmi could have answered few of the boy's constant questions. Now her children were grown and she no longer needed to answer questions, so she smiled down at the boy looking up at her. He laughed and skipped ahead to the edge of the river and started throwing stones at sheets of thin ice, shattering them.

She had paused at the new sheep pen. Seven long, blank faces along the fence stared impassively, plumes of vapor puffing from their wet, black noses. They were still half-wild, but when she touched the noses one by one as she walked by they only tossed their heads a little, as if mildly irritated, and did not try to run. Three yearling lambs stood on the opposite side of the corral, switching their comic tails.

"Lokkani," she called. The boy was named after Lokke, his grandfather. "River's hungry this time of year. Don't tease. Be respectful."

"Yes, yes." He threw another stone, shattering a mirror of ice. The fragments dipped under and were carried away. A bright reflection dazzled her, and she feared for him. It was not good to tempt the river's ire. It would reach for him ... .

Before she could call again, Lokkani was back at her side. "You didn't say," he chided. "When's grandfather coming back? Yesterday you said maybe today and this morning you didn't say anything, so when?"

People were coming out onto their roofs to hang blankets in the sun or work obsidian. Three men stopped by the sheep pen to discuss which yearling they should slaughter for the feast tonight. They spoke together without looking at the old woman and the boy.

Lokkani sulked. His grandmother stared into space and ignored his questions. A family walking to the river with their water baskets greeted her with nods she didn't notice. They were not friends. They had built their house on the other side of the almond trees away from the settlement and kept to themselves. Earlier in the fall they had accused Semmi of collecting too many nuts, and now she remained aloof. Lokkani stared, openly hostile. When he started to speak Semmi stopped him with a tug at the hair on his nape.

"Well?" the boy insisted when the family had moved on to the river.

"Well what?" she asked, distracted.

"Is he coming? It's feast day. He should be here."

"Mmm." She had some trouble catching her breath and sat down on the corner of a nearby half-wall. Someone, she couldn't remember his name, had started to rebuild. But he had stopped when the weather turned cold, and then had left the settlement and had not returned. He had two or three wives, she couldn't remember, but had left with one of them, or perhaps two. And their children. She could almost see his face against the blue above her, beard streaked with gray spreading slowly across a sky growing darker. Clouds? He had a scar down his eyebrow. Or was that the man who lived up river near the hackberry trees?

She shook her head. She could see the girl she had helped bury with her child. Spirit Man had cut a hole to free the bad spirit. Now she had a bad spirit herself, like a rotting thing in her own head. But Spirit Man was gone, buried with his daughters, three of them, bones painted red, when the sickness took them. Twenty years gone at least. Well, she could always summon Spirit Man herself. Of course she could. But it was not certain he would come unless someone with power did the calling, and there was no one in the settlement she could trust with that task. Except Lokke.

Where, she wondered, was he? To the mountain again probably, for the lovely dark gray transparent frozen tears. Or had he gone hunting for the feast tonight? An aurochs, that would be good, a great beast with small glowing eyes and hard hooves to beat the earth. Yes, that would be a good thing. Lokkani would like that.

"Semmi?"

Who called? The boy, was it? Or Lokke, come back at last? She should go home; she should climb the ladder to her roof, go down into her house and lie down there where sunlight sank down the wall and crawled over the floor. Yes, that would be good, to watch the light.

Having forgotten why they had come down to the river, she walked unsteadily back up the road. The boy, annoying as a gnat, ran around her, looking up, calling her name. She tried to brush him away, but her arm wouldn't move, it stuck at her side like a stick of wood.

When at last she was home she sat on the roof with her back against a stack of goatskins. Over the long, wrinkled slope of land to the south she could see the uneven peaks. Swirling patches of dark gray flowed across the sky. The patches came from the mountain. How long had they been there? The sun, sometimes obscuring, sometimes revealing the shimmering peaks, flooded the plastered floor of her roof. Should she go inside? She could lie down in there and watch the sun.

But she was fine enough here. The snow on her roof had disappeared and the plaster was warm under her legs. The mountain! She would keep her eyes on the mountain, streaked with white. More and more these days, people spoke to it; they knew it watched over them.

A darkness like smoke hovered over the white peaks. The mountain reminded her of a hearth, warming her. A ribbon of red ran down its side, so small she could barely see. The mountain bled for her. Then it stopped.

She spoke to it. "Welcome," she said. "I'm glad to be here."

She heard the boy calling her name, and then she no longer heard. She saw his shadow, dancing. She saw the mountain. Welcome, she said, falling into the light. Welcome.

An active volcano was a powerful force in the landscape. Gazing at it would demand an explanation of its behavior that would impose a kind of storytelling order on the world.

Divisions already fragmented the world. There were houses on one side of the road, communal meeting places on the other; there was this side of the river, there was that side. Such divisions could never be undone, just as a core of obsidian once worked into a blade could never return to the state in which the accidents of geology and weather had placed it. Clay baked by fire stubbornly resists returning to clay.

Or, once a bard finished singing a tale of times long past, the song could no more be retrieved than the characters in the story could relive what happened. The earth is forever marked by human activity.

A well-traveled track connects distant communities. The days when mountain peaks, significant bodies of water, rivers, or transient things like trees could serve as primary focal points of navigation in a disordered world were passing away.

A track cuts the landscape in two and divides wild land from itself. It discards the two parts until they ready for exploitation for agriculture or pasturage. Only the point of departure, the destination, and the line between them remain.

Thus the taming of wild nature began. Like the stone tool, it would never go back.

As far as the mind could see, tracks crisscrossed the land with names and traditions of their own. Over them ideas flowed in all directions.

Square houses built side by side hastened the smoothing and regulating of an unruly world.

The Journey

Once abandoned, the settlement will fall into ruin and be buried by the years. Some people will die with it. Others, attracted by curiosity for what's over the horizon or driven by adversity, will leave. The result is the same. The forces of the natural world were always present, but as populations mix, ideas incubate, technologies evolve, social structures intersect, and cultures change, those forces recede from mankind's attention. Eventually, for all their power of life and death, they would become distant gods.

These are the new journeys of mankind, to move not only through space, but also through our past.

Lokke kneeled beside Semmi, his woman for twenty-six years.

The boy Lokkani was there, looking in continued wonder at his grandmother's smile. Her eyes were closed. "She went to sleep," he said. "She came up here and sat and looked at the twin mountain. I could see something over it, too, a darkness. Can you see it? She looked a long time," he repeated, shaking his long hair across his face, hiding his large dark eyes. "But she was smiling. Then she went to sleep."

"Yes," Lokke answered, taking her lifeless hand. "I know."

Lokkani had been waiting for her to wake, but now, as the sun neared the land across the river, he understood she would never open her eyes again. He regretted throwing stones, taunting the river. She had warned him it would be angry, and now it had taken her from him.

Looking out over the settlement Lokke understood for the first time how this place had changed since he and Semmi had made their children.

For years people had been leaving, a few at a time, and he hadn't really noticed how their absence changed things. Some simply left in the night without a word to anyone. Some said they had relations to the east or the west and were going to join them. Others, the angry, talkative ones, complained that the land had turned against them: grains were scarce, and even when they put them in the ground and tended them they came up thin and weak. And the weather was colder than before, and the rains came late, or too soon, or didn't come at all.

The community was fading like the day itself. So soon! Lokke thought that like Semmi lying here, her back against a pile of skins, it would soon be gone.

Today was Feast Day! Though this was the shortest and coldest day of the season and tomorrow would be longer and the sun would begin moving toward them once more, he knew today marked an end, for him, if not for the settlement. He had only to look at the mountains gleaming white with winter snow, somehow dimmed by shadow to know it could not endure.

The settlement was dying, this was true: the neighboring house had already fallen in upon itself, and the house at one end of their small neighborhood had recently been abandoned. Each year the Feast Day gathering was smaller than the year before. Soon no one would come.

Could he trust his memory? He had long passed his fortieth winter and had frost in his beard. Perhaps he was imagining it all. It was just his grief and loss, his shifting mood?

No, too many houses were ghosts never properly buried.

He looked at her, the mother of his two children. He was sure they were his, after all. Not every man in the settlement could say he knew such a thing, but he could say it. They had looked much like him, with the same dark eyes and broad forehead.

They, too, were gone.

He should have known. He had never been good at seeing ahead into the round of time, but when he was at the mountain of frozen tears just two days ago, only three fur trappers had appeared from the east to trade. Normally there were ten or more. Every year they spoke of their lands to the east, of the Great Stone Circles near a mighty river. "They were rebuilt this year," one of them told him just a few days ago. "The pillars look down on humans; they are powerful. Some of our people helped build them, some even saw the secret ceremonies. They told of fire and animal companions from out of the stone. My nephew was taken there when he had twelve winters, and when he came back he was much changed. Great power is there, my friend, even after generations without number. Great power."

"Then why are you so few this year?" Lokke wanted to know.

The trapper only laughed and showed him a fine wolf pelt.

Lokke exchanged one of his best nodules of frozen tears for it. He would give it to Semmi as a cloak.

On the way home, though, he came across a raven plucking the eye from a man dead on the trail, and had felt a chill. Yes, he should have seen this. All that was familiar was already gone. Semmi was dead and there was little left here to trade for his frozen tears.

There would be food down in the house. Semmi always kept sufficient supplies, and there was the feast tonight. Tomorrow he would carry her to Twin Mountains. She had died looking at them. They all looked at them, the most important mark in the circle of their world, but the look on her face was different. She *knew* the mountains.

Semmi had seldom shown fear, but after her daughter died giving life to Lokkani she had changed. Spirit Man could do nothing. Five years ago Sun Demon took their younger son and Spirit Man had been helpless, and just two days after that, Sun Demon took Spirit Man himself along with his three daughters.

They were all helpless; the cords that held the community together were coming undone. Since the deaths, first of their son, then Spirit Man and his family, Semmi and some of the other women had started speaking to the mountains. She told him just before he left for the frozen tears that the mountains answered her.

He cleared his throat. "Come, Lokkani," he said, standing up. "It is time for living; this is Feast Day."

"What of Semmi? Where has she gone? Is she with Uncle, and with Mother?"

Lokke did not answer. What could he say?

Neither of them felt like eating, but even the boy knew they could not avoid attending Feast Day. He sensed somehow this was would be his last.

"The cold will hold her. Tomorrow we will carry her to the mountains. The darkness above them tells us they have roused into life. I have heard tales … . She told me it spoke to her, the mountain. It called her. We must answer by bringing her to it."

So it was. Next day a crystal sun gleamed across a sky the same milky sapphire as a stone someone brought through years before. Lokke went from house to house to find those who would go with him. Only Lokkani's father Tarn, along with his sister, the one whose man had left two winters ago when their child was born, agreed to leave with them. Lokkani's sister would carry her son on her back though he had two winters already and would be heavy. "There is nothing more for us here," she said, already gathering sacks of water and grain, dried fruit, nuts, and strips of smoked meat tough as leather to carry with them. "This will last us seven, perhaps eight days. The mountains decide if we find something more after that." She did not seem disturbed by the uncertainty. Her brother was a strong hunter and Lokke had frozen tears to trade.

The three adults and the boy walked for two cold days separated by a colder night. Near the highest peak the air cut like a blade of frozen tears and a strong smell tickled the nose and turned their heads away in disgust.

Lokke had visited this place only once before and found its frozen tears inferior, small and easily fractured. Its peaks had always been steady through the years until yesterday. Yesterday it began making silent signs dark against the sky, calling to them, though Lokke could not say whether in anger or command.

They toiled up a long slope. It had become strangely warm, and high clouds veiled the sky. The steeper sides above them were covered with snow streaked with gray and black, but the foothills were bare. Sharp edges and spars of black rock thrust up and turned their feet. The smell grew more intense, burning their throats. They picked their way around the western peak and stopped on a nearly level space beside a jagged ravine.

With a sigh, part regret and part relief, Lokke put down the crude travois on which he had been dragging Semmi's body, tightly wrapped in the wolf skin cloak she had never seen or touched while living. A cloud of smoke flew upward into the darkening sky and the mountain roared. It was angry and pawed the ground until it shook. A small spot of red appeared in the gloom and crawled slowly downward, stretching into a thin line that narrowed and disappeared.

He closed his eyes and watched the green line fade against his lids: life, death, new growth. Spring was coming.

"We're tired," he said to the others. "We stay here the night. The mountain is glad we have brought Semmi here."

"There's no water," Tarn complained. "Nor is there shelter."

"It will soon be dark and both water and shelter are far. It is almost flat here. We leave at first light."

Lokkani was unhappy and silent, all his questions stilled. His grandmother's face lay peaceful beneath the somber sky. He told himself she was glad to be here, as if her smile were even happier.

They huddled together. With no wood their meal was cold and meager. They shivered in their furs. The child cried intermittently but soon caught the mood of the others and lapsed into silence of his own.

Just before dark a low rumbling began to sway the ground. This time the movement was almost gentle, as if they were babes in their mothers' arms, and for a moment Lokke thought perhaps the mountain's anger had passed. Lokkani cried out when the arms of Semmi's crude travois lifted into the air and it jittered over the edge of the ravine.

They watched in helpless horror as the wooden frame splintered, sending broken poles and cross beams spinning away. They could see only the gray wolf fur and Semmi's dark hair as she tumbled to a stop just above the darkness at the bottom.

Far below, she faced the distant plains as though contemplating the sun now almost gone under the earth.

Lokke straightened. "Tomorrow we'll leave the Twin Mountains and go in the direction of her gaze until the land asks us to stop."

The shaking grew more violent toward morning. Smoke rolled down the slopes, enclosing their camp in a black and gray cloud filled with floating ash, an oppressive smell, and the taste of burning. They fled in the darkness past the mouth of the ravine and onto the open plain. With no stars to light their way they clung grimly to one another. The mountain's grumbling, interrupted two or three times by loud thuds, slowly faded as they ran.

Lokke led the way. A strange fire had possessed him, burning in the impenetrable darkness with a fierce black flame. On flat ground away from the mountains they slowed their flight. For hours they marched steadily over gray spongy ground fractured by twisted fissures and mounds. When they stopped that evening, Lokkani, seated cross-legged on the ground, took a sharp stone and scratched squares and ovals and crude animal heads into the ashy, friable earth until the light was gone.

On the afternoon of the second day snow began to fall and they had to stop. The next morning they managed to gather enough to refill their water skins, but their food supplies were almost exhausted. They set out early in bitter cold. Lokke regretted leaving the wolf skin with Semmi but quickly set his doubts aside. Surely by now the mountain had taken her into its fiery heart and the wolf's life would cloak her in all its fierceness. She would surely need such intensity where she was going.

That night the clouds lifted and sharp stars leaped out, bright enough to cast shadows around the endless spires of stone. The next day turned unseasonably warm and from time to time the ground shook. In places it became soft and treacherous under their feet. Then all was calm and still. It seemed the mountain was reluctantly going back to sleep.

At midday they came across a hut beside a small stream. The five people there spoke nonsense, but after extensive gesturing Lokke managed to trade one of his obsidian cores for two sacks of wild grains.

Warm weather continued the next day. The ground grew increasingly marshy, and they had to pick their way around pools of muddy water and over fallen limbs and clumps of last year's vegetation. Animal tracks winding in long-worn patterns helped. Late in the day Lokkani's father outlined a huge hoof print with his finger. "Aurochs," he said, pointing the direction they were going, to the left of the setting sun. "Five, probably a family." He hefted his spear, a puny gesture in all that emptiness. The aurochs was an enormous animal, fierce, ill-tempered, and dangerous. A single man, even with help from an elder, could not possibly kill one. They should avoid confrontation.

The next morning they discovered their best intentions were easily thwarted. They heard shouts ahead, a man's voice soon joined by others and the ground shook as it had on the mountain. Hooves approached and a panicked family of aurochs, perhaps the same that had passed this way, now turned back, burst from the low, dry shrubs, and rushed toward them.

The trail was narrow. Lokke grabbed the boy and pulled him into the swamp. "Run!" he shouted, but the others had already plunged off the trail.

The boy's father turned back and hefted his spear.

The lead cow rushed toward them, eyes wild, swinging her horns from side to side. Several spears flapped from her sides. Blood flowed glistening along her flanks. She ignored the man standing by the path and when his spear plunged into her brisket, she heaved a great bellowing sigh. A sinuous thread of foam-flecked blood flew back from her mouth and lolling tongue as she ran past, followed by the others.

The hunters, wearing strangely fashioned leather caps, ran easily some steps behind. One stopped. "You threw," he said to Tarn. "I saw." His words were curiously formed but clear.

"Yes."

"The cow is wounded."

Tarn nodded. "She will not last long."

"Come with." The man sprinted away. Tarn did not hesitate.

Lokke led the others back to the trail. "We wait."

The sun was nearly overhead when Tarn and the others reappeared carrying large cuts of meat. Tarn presented his family to the hunters. "These are my people," he said. "Lokke is our elder, as you can see."

The hunters, seven of them, greeted Lokke and the rest with respect. All lingered to share food and talk of the hunt. The strangers praised Tarn for his skill and courage.

Lokke asked them where they passed their days when they were not hunting. They replied that the land toward the setting sun was rich with game and wild plants, and some families had stopped there. "Do you build houses?" Lokke asked, but they did not understand him and shook their heads.

The hunter who had spoken first to Tarn stood. "I am called Sharzak. We, too, see the darkness over the mountains. We honor her, our healer and guide. She speaks to us as you say she spoke to your woman, whom you have given to her. That is good. You say you left your lands and will not return, and that, too, is wise. You come with us, for we are few, and you may find our land to your liking."

So they followed Sharzak and his men. After another night they reached a broad meadow, winter brown with dry crackling grass beside a river, smaller than the one at home but as graceful. All around were thawing wetlands. Open space, rich in low shrubs but heartless in its emptiness, stretched in all directions to an uneven rim of distant mountains. When they turned back toward the rising sun, they could clearly see the two peaks, which Sharzak called "Her."

There were eleven stout hunters and their women and children in this settlement of skin tents and brushwood walls huddled near the riverbank, thirty people in all. Smoke from their fires rose straight into the still air. The sun shone brightly, and it remained unseasonably mild.

Sharzak said they now remained here through the round of seasons, winter and summer. "We have no need to pursue the animals," he said. "Goats find plenty to eat and don't go far as they used to. The land is rich. In two or three moons there will be plenty to eat for the rest of the turning of the seasons. Stay with us. We can hunt together, and your stone, your frozen tears as you call it, cuts well."

Sharzak offered them baskets of grain, fruit, and nuts. Most curious though were their slabs of salted meat. "What is this taste?" Lokkani said, wrinkling his nose at the first bite.

Sharzak explained that the bitter waters to the south changed the meat so it could last for a long time. They traded wild grain and goat meat for the white crystalline stuff they could cure the meat of goat and sheep, as well as wild animals they trapped or hunted.

"We smoke our meats," Lokke told them. "Never have I tasted something like this."

Later he asked how many rounds of the seasons they had remained here. Sharzak began to count on his fingers with his face screwed up in such a comical way that Lokkani had to stifle a laugh. "Seven," he said at last, pointing to a young girl. "It was the year she was born that we first stopped here. This is her eighth winter."

"You have stayed so long in tents and shelters quickly built and in need of constant repair?"

"Of course, why?" Sharzak asked.

Lokke described how they could use long loaves of clay dried in the sun to build walls that kept out the cold and held in the heat in winter. "People in the lands toward the rising sun have done this since the Beginning," Lokke said. "We did it for many cycles of men."

"We have heard stories of such things," Sharzak said. "But so much work! We have little time to spare for such foolishness. We find seeds and fruits, and we hunt, mostly goats and sheep. Sometimes, like today, we kill an aurochs and then we have a great feast, as we will tonight." He shook his head and grinned. "We have time to sing, to dance, to play games. Why should we do other than such things as give us pleasure?"

"Once you build you don't have to build again for many years."

"This leaves us more time for play?"

"More time for play," Lokke agreed, but Sharzak was doubtful.

Semmi's mountain, a dark outline against the rising sun, went back to sleep. The black cloud drifted away. Lokke said this was a sign the mountains were permitting them to remain with Sharzak and his people in this well-watered place.

When it was warm enough to dig clay out of the riverbed, Lokke and Tarn began to build a house. The others watched curiously. Sharzak mocked them, saying they had no time for play. But finally when the house was finished and the weather had turned warm, he had to admit that it remained cool and pleasant inside.

When the hot season was at its fiercest, some of the others began to build as well. Instead of leaving small spaces between the houses as Lokke's people had done, they built right against their neighbors.

In the frigid depths of the following cold seasons, fires inside the houses made cheerful warmth, and all who lived there were crossing roofs and literally dropping in on one another.

Thus began Çatalhöyük, an important World Heritage settlement on Turkey's Konya plain. The small percentage uncovered by archaeologists so far has revealed details of its long occupation.

Many still lived by foraging and hunting. Nomads still tended docile sheep and goats, leading them back and forth from summer to winter pasturage, sometimes over great distances, but most humans were no longer following the food, as in the Paleolithic; the food was staying close to home.

Those who settled permanently kept their animals close. They devoted more and more time to cultivating grains and other crops. They shaped objects out of clay like bricks and figurines and small balls they could use to heat water for stews and soups.

The settlement's basic pattern—sun-dried mud brick walls, roof entrances, hearths, storage bins, and sleeping platforms would repeat for hundreds of years across thousands of houses.

The new factor was density. The dwellings pressed against one another, with no space between them except accidental gaps where houses failed to connect. In effect, they became underground caves. The dead buried under the floors were the underground of the underground. Although house burials had been practiced for a long time, crowding pressed everyone into greater intimacy with death.

Çatalhöyük is still not a town. There are few if any public buildings or public spaces, no marked differences in the buildings. Most people probably spent time away in the warm season tending their animals or plots of cultivated land.

When they were in residence those double thick walls were a second skin around the bodies of those within, but protective walls can also isolate.

Mud People

The new community grew from the center, house by house. Together the roofs above them became their social spaces that left no record.

Domesticated sheep and goats lived much of the time in odd-shaped spaces between houses also used for trash. For more than a thousand years the agglomeration grew, until it became a hill above the plain, a permanent landmark.

Thousands of residents built here, for it was well-watered and rich in fowl, fish, fruits, and grains. Not too far away were places to farm, and farther away, wood for building. Above all, there was clay.

On a crisp morning in late fall, four men brought a body wrapped in a blanket from a house near the center. They moved awkwardly toward the edge, crossing one roof after another. The frost of their breath blended with the smoke of cooking fires, for it was winter cold. The morning star gleamed blade-point bright over the double peaks of the mountain where Lokke and the others had left Semmi. The star faded as the rising sun threw long orange beams into their eyes from beneath layers of striped gray cloud.

When a curious observer asked, they described how the dead man had fallen from his cousin's roof into his neighbor's animal pen and broken his neck. Now, as the cold season had begun, he must join others important to him and his community under his uncle's house. They spoke in short, breathless bursts.

"Who was he?" a passing woman asked. No one could know all the people who lived in this place, and every year there were more.

"He was Tarsh, but his name no longer matters," the oldest replied. "He fell and his name left him. We return him to Alazne, his uncle, a Keeper; he is responsible for the dead of his clan."

The others nodded. "We did not know Tarsh, but we know Alazne." Everyone, it seemed, knew Alazne.

They picked up the body again and struggled on. Lugging the body over the uneven terrain and up and down ladders was hard, but at last they reached Alazne's house near the western boundary just three roofs in from the outer edge.

They laid the man who had been Tarsh on the roof. The oldest called down through the smoke hole. "Alazne! We've brought the young one."

The old man struggled up the ladder. He stood over the broken man and chewed on his lower lip. "He was my nephew," he said in a quiet voice. "He had skill shaping obsidian into blades. He would gossip as we worked. We made a kind of family, you know, bound by our common skill. All would know him if he had lived."

The eldest agreed. "He deserves a place among the makers under the floor."

Alazne said, "Their bones are echo and breath of life. He will join them in the mud and clay."

When conversation faded, they lowered him into the dark warmth of the interior. The oven was already ablaze and two women of Owl clan, his totem, were heating a stew of freshwater shellfish and wild vegetables. The young man looked on, growing cold and stiff, while they cooked.

Neighbors, drawn by the smells of food, were already telling stories about him, his reputation. Those who lived closest gathered on nearby roofs. Some anticipated a celebration, for always with a death came food. Better yet, there would be ample drink of fermented goat milk. The dead always demanded the honor of food and song. This now nameless one was the first to be laid among this nest of houses for several moons. His death was a chance to stop working and remember, a time for sorrow and its passing.

The small space was crowded, but there was room in the center for Alazne. He dipped a branch of tamarisk into a wooden bowl and shook droplets around the room. His circle ended at the skull of a bull set into a bench, watched over by a plaster relief of a bear splayed against the wall.

Alazne had often told his nephew the about the bear. Now he told it to the people crowded in this house, though they had heard it many times before. "It came," he said, "from a vision one winter long ago when I was almost a man. I dreamed I had killed it and spread its pelt to dry on the roof. So I shaped the bear in plaster, and it has lived in this wall for many years."

The horned aurochs skull was another story. "A handful of years ago one of the bull's horns poked out of a bog two days' walk south of here. The bull adopted me. It was awkward to carry, and it took more than half a moon's cycle to bring the skull here, to bring it over the roofs to his house, to lower it into the house, to make it part of the bench. But Bear and Bull give words when I, Keeper of the dead for Owl clan, need them, as now."

He stood in the center, breathing hard. The silence held as his breathing slowed. He began a low, nearly tuneless melody. As it grew louder, it began to take a shape in the listeners' minds. Repetitions and modulations emerged, and the audience turned to one another.

Alazne was invoking the house itself. He sang of its founding, when the first people made the bricks of mud and laid them in lines and turned the lines at the corners. The corners became a square of walls, and these walls separated neighbor from neighbor, clean inside from dirty outside, what was known only to the family and what was known to all.

Once, Alazne sang, the house was at the edge. Now it was three roofs in. This was the sixth house in the generations of houses built one upon another, and beneath six generations of floors dreamed the bones of ancestors, the bones of bonded brothers, the bones of nameless infants and lost children.

His song recalled how many roofs the dead man needed to cross to reach his own. As Alazne sang, he scattered more droplets of river water on the sleeping platforms, the plastered floor, and the low round entrances to the storage bins to the north.

Lifting his arms he chanted roof beams into place, and thatch and mud to seal it. The house was the family's body. There was darkness within, an open mouth in the roof. The house was a living being. The smoke and people were the breath that flowed in and out, up and down, through the opening. The house was a being, born out of its past and the people of its past, and the bones of the house were brick, and its sinews were reed and wood, and the skin of the house was lakebed marl and mud plaster.

When he straightened at last, his knees cracked. "Now is the time to lay this body in the platform against the east wall. Winter is soon here. Already cold grips the waters and we will remain inside. It is fitting that this man goes into the earth below, for this house, with cattle skull and bear relief and red palm prints on the wall, is a living body."

The visitors murmured agreement. They drank the fermented goat milk, and many fell asleep where they were. The fire in the oven died down to embers, and outside darkness fell.

The young man, who had been called Tarsh, was sent from his grave into the dreams of those sleeping above, all that he had been and all that he had known.

The dead disappeared under the floors where they became the non-ordinary spirits. They would continue, through memory, to give meaning to the living until those who knew them also died. When someone new was ready for burial and old bones were exposed in a fresh grave, an elder might recall stories of the dead, even those whose names were lost, and recount what could be brought up from the buried darkness of memory. Thus the house was a cache of ancestral memory and meaning.

In the mythology of the Maya Indians, the gods first created men out of mud, but the mud people were not viable because they couldn't speak or move and soon crumbled.

At Çatalhöyük, mud was fundamental. Stone was scarce, but mud and clay were plentiful. Mud filled their pores and flowed through their veins. It was the bone and muscle of their houses, the enduring, constantly renewed surface of their walls and floors.

These were the Mud People, and unlike the Maya (who settled on maize as their substance of origin), they could speak and move and build.

Like cave walls in Europe, the white plastered interiors of their houses invited a hand to recreate the world outside. Aurochs skulls mounted in the walls or in rows along the floor, relief plaster sculptures of bears, walls embedded with vulture skulls, or paintings of vultures with spread wings decorated some of the houses. Like family photographs, they prompted stories of those who lived before, who they were, and the deeds that gave them fame.

Like caves, these houses allowed only limited access. Each was built in the same basic pattern, around twenty-five meters square, a modern studio apartment. The ladder down from the roof on the southeast created a symbolic vertical journey. The ladder ended near the oven. Smoke thickened the air and dimmed the light, though freshly plastered white walls could dispel some of the gloom. Two or three sleeping platforms a few centimeters high lined the walls to the north.

In one late house, inward-facing aurochs skulls formed two sides of a small corner platform. An infant burial underneath suggests their purpose was to protect the small body.

Motherbaby

Some houses enclosed many burials. Sometimes the dead were relatives, sometimes not. Sometimes, as with the Snail Creek Shaman, the dead would prompt a story.

When the young woman died in childbirth over seven millennia ago, she was buried under the sleeping platform, as was the norm for her time.

Aside from her burial, though, there was nothing normal about the motherbaby.

She was called Semnali, a variant of an ancient name among the People, she of the Twin Mountain clan, for her lineage came from the peaks visible far to the northeast. These women knew well that their task was to care for her. Should Semnali fail to release her child into the world they would take the fetus and bury it in the south part of the house near the hearth. Should the mother die, they would bury her tenderly under the platform and plaster it over.

They would do the same for anyone, but today this was a special case, and they knew that someone, perhaps one of them, would return to manipulate her bones. Since memory began this was an oft-repeated drama.

What made this day different was not just her place in the community and in an important lineage, but the way she met the challenge of this difficult birth. The way she paused at the threshold, ready to go either way, and then made her choice, the most difficult of all.

The morning her waters broke she was scraping the inside of the pelt of a wild ass, working it into soft leather.

Though Semnali made no outcry, the women appeared as if summoned by some secret signal. They helped her down the ladder and laid her on a thick bearskin on the eastern sleeping platform.

She was a beauty. Everyone said so. Her flawless skin glowed from within like precious amber, her dark eyes sparkled, and she carried the weight of her growing child with ease.

After some time of normal breathing, though, her pain, a predator watching for weakness, pounced. Her distended body thrashed violently, for she had become something utterly elemental, a chaos of groans and excrement and blood amid a stolid group of elder women murmuring soft words. If they patted cool water on her face or moved her into a more comfortable position, she scraped at them with her nails and begged them to make it stop. This was her first child and she had no prior acquaintance with the endurance she must summon. The women held her arms and massaged her belly. They could feel her heat. "Semnali," the women crooned, brushing her hair away from her forehead. "Semnali, Semnali, there now. The sheep in the pen, the goat in the pasture, they let their babies come. Be like them and let it come."

Hours went by, and she tried to let it come. She cried out and pushed and gasped for breath and pushed again. Her teeth flashed between twisted lips. Her forehead beaded and flowed, and still she cried and gasped and clenched her body until it shook. Her hands tore at the bearskin, pulled out fur by the handful. The women's soothing words were swept away by the torrent of her pain.

Overhead men hurried across the roof, uneasy at the shrieks from the entrance. Other women on neighboring roofs whispered to one another of her beauty, the grace and strength her baby would bring to the community, the fine deeds he would perform or the gifts she would bring into the world to bestow upon others. It was important to welcome a newcomer into the community, for its future depended on them, small, helpless beings that would grow into integral parts of the whole of the People.

Down below she was begging for the pain to stop. She cried for help, but there was no help for her. A force stronger than she, than the women attending her, the whole community, took hold and would not let go. This force, the women knew, was the origin of the world, the way the world always formed out of the darkness in agony and blood.

They had seen it before, had known childbirth themselves, knew how the large head seemed to split a woman open like a living cocoon, how the creature inside twisted and kicked until it seemed it was either fighting its way out or digging in its heels and refusing to come despite the mother's overwhelming urge to push it from her.

Semnali's cries grew fainter, farther apart. She panted short, shallow breaths. Sweat blinded her, pasting her thick hair to her temples. This baby, *her* baby, would not come. It would *not*.

Instead blood poured in a sudden gushing tide of crimson life between her legs. The elders wiped it away, still murmuring between her harsh indrawn breaths and feeble shrieks cut short. They turned her this way and that. They exchanged glances, and as the hours passed the glances became more somber.

No one was surprised when the screaming stopped.

The flickering dung fire in the oven exhaled a faint reddish glow and sweet-smelling smoke. Through the roof opening afternoon sun highlighted the elaborate headdresses of the five women around Semnali and the child trapped in her cooling body.

Her head was thrown back, her teeth bared. Now, as her face slowly relaxed, someone tipped her head forward and closed her eyes. She was at the very edge of the great water, of the vast dark, of the gulf of memory. She, and the child. At the edge.

No one would know if the child was a girl who would one day have helped the next wave of People into the place they called Ju-Ni, in the midst of teeming, wet life, or if it would be a manchild who would shape the next turn of the house and so add to the growing mound, the husk of a once-living creature. When her bones were next uncovered all flesh would have disappeared, the child's identity erased forever.

There was a reason the child, girl or boy, had refused to come, had declined to separate. Motherbaby slipped from pain to peace here on the recently plastered platform in the midst of all the houses of Ju-Ni. They had gone together to the edge, and had turned away from the world of the living at the last. The ancestors decreed it was meant to be thus, that her sacrifice, their sacrifice, was necessary.

So these women would sing the Song of Farewell. They would make the other sounds, too, the drumming and the high bird flute that would carry Semnali into the underfloor dark.

She had come to this house three years before, when it was newly built. Together as motherbaby they would remain beneath it. Her/ their life would pulse up though the walls, old and new. Should these women stumble during the ceremony, motherbaby could become a vengeful cause of breaks in the roof thatch, dropped utensils, broken tools, or sudden surges of anger. But if the telling, the chanting, the manipulation of her soft flesh, the washing of it, covering of it with perfumed oil or red ocher, and the careful placement of objects with her in the underfloor were done according to custom and with attention, Semnali and her unborn would incite unprovoked smiles or surges of joy and the house would flourish. Turning like this as a motherbaby at the edge was a serious thing and demanded much of the living.

This was the way it had always been: only Ju-Ni remained, growing slowly toward the sky as the People came up into the light and went down again—living into light, dead into dark, over and over.

The wise women stroked her soft, young shoulders and cheeks with their fingertips. They took strict turns telling of the way she laughed, the clay figurine of a bear she had kept with her all her years, and the young man she had lain with, the father of her child. They spoke, too, of her skill at making flat bread, how her mother had taught it to her as her mother before her had taught her, and how she had told her mother when she was first named that she would do the same for her child, girl or boy, and how it made her mother laugh to think that a boy would learn to make flat bread!

They spoke of Semnali, and they spoke of the other part of her, an essential part still nameless, not yet a person and so a mystery, a potential. What might it, after it came into the light and announced itself, have become? They asserted in their high, singsong voices that it would have grown strong beside the oven, and, named at last, joined the People. It would have been manifest, flesh and bone. Now it was something else, something more, a seed, a green shoot of winter wheat with great power to help the house endure and flourish.

Beneath this new house were the bodies of other, earlier houses with bones of clay interlaced with the bones of their people grown together into a durable matrix. Up here in the semi-light the plaster was fresh, the thatch still smelled of wood and field and the oven cast warm shadows into the room. The floors and walls had the texture of a just-born, its soft white surfaces pleasing to the touch. This was her house. Now that she was no longer living, she/they, motherbaby, would be the first to be woven into its fabric.

Two of the women took obsidian blades from a low-rimmed basket and began to carve a large spiral out from the center. Over and over they cut through the plaster and into the dirt below. They peeled up the white spiral and set it aside. They scraped at the dirt and piled it beside the pit in neat mounds, releasing the smell of damp and decay into the chamber.

While they worked, the others wrapped the dead in a green cloak and folded her into a crouching position so that she made a circle of her own. She was now a little smaller than the open pit on the platform.

Several women and two men entered now and moved to their places around the perimeter of the room. They pressed back against the walls and looked on with impassive faces. Only the awe in their eyes fixed on the dead woman's swollen belly revealed what they were feeling. They were the witnesses, the mother and brothers and cousins, the sisters and those with the most rigorous ceremonial ties. Many died in childbirth, mother and child, but few crossed together.

The two diggers finished preparing the pit in the ground and sat back. The eldest, who was Semnali's only living foremother, crouched beside the motherbaby body and lowered her head and said, "Semnali came in blood and pain from the earth that sustains us. She grew inside the circle of our walls. She lived in the shell, the rind, the husk of her house. She grew an unborn in the fern of her womb, the hollow of her belly, the house of her body; she is one, they are one, they are two. Together they go now back to the mud, together the mud takes them in; together they flow back into the lives above. They follow us; they lead us; they are us."

The others murmured agreement with nods and tongue clicks. "Yes, yes," they repeated. "So be it. So it is."

The oven fire flickered. The light from the opening above lost its brilliance and dimmed as though a cloud had passed across the sun. It was late in the day.

"They hover like the hummingbird between this world and the other," Foremother intoned, her elbows down at her sides, wrists bent back and palms facing up. "*From* the life of the house *to* the life of the house, *now* is the balance. Firelight and sunlight are equal; human and not-human are one. This woman was Semnali. She and her Not-Yet precede us. Soon or late we follow should the People deem it fitting. Now, at the meeting of fire we made with sun that made us, we return this hidden seed in its shell through this opening in the roof of earth."

One of the women handed Foremother a funeral basket, the kind used to bury stillborns or neonates. The young dead were revered, and because they were so numerous, they were buried in the house as a continuous living part of it. A basket like this, lovingly woven, was an honor and an offering of thanks. Its repeating design of green interlocking spirals against a cream background defined its purpose: new life at the center, enclosed by mother life, house life, community life, the wide world; new death at the center, protected by mother death, house wall, community wall, and the great, uneven horizon.

Foremother carefully placed the basket on the floor of the shallow grave and there was for a moment mingled with earth-breath a stronger smell of green things, of new-cut grasses oozing juice.

Four of them lowered the motherbaby into the pit on her right side so her distended belly lay cushioned on the basket. They arranged Semnali's hands by her knees with the palms bent back in the gesture everyone used evoke entities of the two skies, day sky and night sky. Her head was bent forward so she could look at this growth within her, so she could speak with it, so they could talk together, mother and child.

"For her, this woman, Semnali, and it, her child," Foremother intoned, taking up a bone pin and carefully placing it on the dead woman's stomach. It was the pin she would have used to hold her cloak together, the one she would have used to swaddle her infant. Foremother covered it with green pigment, the green of the basket beneath the woman's belly, the green of a shoot of winter wheat.

Grandmothers in many cultures were the midwives. Aiding in this difficult passage is one important reason human females, unlike most other mammals, live long past their childbearing years.

After the children are born, grandmothers become the primary minders of children while the young mothers forage for food. This is called the Grandmother Hypothesis.

In this story older women also act as midwives to the death, which is in its way a birth in reverse.

How the living perform burials—prepare the body, arrange it, and what they place with it—shows how a culture manages memory. Time itself was a rapidly changing concept; the long cycle of hunting and gathering was giving way to the dominance of linear time. Without the enduring traces of previous generations, our ancestors had little sense of a past deeper than a generation or two. There were no headstones, no Bibles with the family genealogy inscribed in them, no history books noting contributions to the advancement of mankind.

When a house reached the end of its useful life, they carefully dismantled it, buried it, and built a new one on the stubs of the old walls. The ostensible reason might be structural integrity. With no local stone for foundations, a house was carefully dismantled and buried like a deceased person. In effect, the house was dead. Emptying it, knocking down the walls, crushing the old bricks and cleaning the dust, spreading it and compacting it to make a new floor, all prepared the house for burial like a body. Building on the stubs of the old walls resurrected it.

Sanitizing and brightening the interiors by spreading plaster and polishing the surface with a smooth stone was a performance that drove into living flesh a deep kinesthetic sense of the continuity of shared history.

The dead were remembered. Descendants slept in close proximity to them, and when someone else died and they dug under the platform for burial, older lives came up into light once more.

Plastering over her grave was not the end of motherbaby's story.

The Beast Without

Many animals, particularly sheep and goats, were living in close proximity with men, but one, the aurochs, remained wild and unpredictable and terrifying. This is one reason their skulls held such power inside the house. Installing the skulls was an initial step in the aurochs' domestication.

In a community of domesticated animals and increasingly inscribed land, such a fearsome natural force could not long remain untamed.

On the morning of Semnali's labors, Mánus of Waterbird clan, the unborn's father, was watching Foremother dip a stamp of sun-hardened clay into a low stone dish of moistened charcoal and press it onto a small blanket of coarse goat hair. He couldn't tear his eyes away. Black lines in the shape of a looping stylized bear leapt vividly from the off-white background when she lifted the stamp away. He was entranced by the way the image appeared each time until the animals marched one by one along an edge to the end.

The blanket was for the person-to-come. When he or she finally separated and entered into Ju-Ni, the women would swaddle it in this blanket.

"Here," he said, presenting her with the back of his hand.

With a grin she dipped the stamp and pressed it onto his skin. He looked at the nested loops that made up the bear's outline, its small ears, splayed feet and plump belly. It was an autumn bear, fattened and ready for winter sleep. The child would grow up with this line of bears clutched in its tiny hands, and for a time now he, too, had one. This was something they would share.

He glanced at Semnali. She sat cross-legged on the next roof scraping carefully at a hide, her head down, intent and focused. For a moment he resented it that she never looked over at him, and he laughed at himself. How foolish he could be sometimes! She was going to have a child.

She stood suddenly, eyes fixed on the plastered surface between her feet, and as if in response to some unseen signal, four women flowed around her. He couldn't see what Semnali was looking at, even after the women had escorted her to the entrance.

Semnali's foremother thrust the blanket into his hands. "Take this," she rasped. "Hang it up to dry." She was over the low wall that separated their houses and inside his before he could answer.

Many times he had been nearby when children were born and always the women sent him away. When his sister and both his brothers separated, his father, too, was excluded. There was something mysterious and terrifying to men about this birthing, something the women would not explain to him no matter how often he demanded answers.

He draped the blanket over a wooden rack. The bears seemed to be cut out of the cloth, as if he were looking through the blanket at a darkness underground. The bears were shambling in a line, one after the other, off the edge.

He shook his head. For a moment he didn't know what to do, only that he had to move.

The house he shared with Semnali was closed to him. Abruptly he scrambled over a series of roofs. "Cousin," he shouted into an opening. "Loan me spears. And some rope."

A round head appeared through the entrance. "Mánus? Why do you want spears?"

Mánus shook with impatience. "Just give them to me. Semnali is birthing and I have to go. You'll get them back."

With a grunt his cousin dropped down into the darkness. He was a fat, lazy donkey of a man, a bit older than Mánus, and no hunter. He would rather sit on the roof drinking sheep's milk with the women than do anything else.

In a moment he was back. He proffered the spears and rope, his mouth turned down in a sour frown. "Take."

"I'll return them."

"Don't bother. I don't need them."

No matter how far Mánus ran from the settlement, he could still hear Semnali crying out. Of course that was impossible, she hadn't made a sound before he left and he was soon too far away to hear anything. He was just remembering those other births he had been near. His pace slackened to a trot, then to a walk.

Here he was, a coil of rope in one hand, and two long, slender spears tipped with wicked obsidian points in the other. His cousin had not made these points; he had no real talents except eating and gossip.

Even the small stand of wheat he occasionally tended produced little grain.

Mánus walked along a game trail beside the river, passing signs of deer, rabbits, wild dogs, and once the scat of an elusive leopard. To his right the land sloped up gently and leveled out. Clumps of tamarisk, hackberry trees, and shrubs of a hundred shades of green dotted the plain. Animal life teemed on both sides of the river. He could see a small herd of deer, quite far away, moving slowly against the purple-gray of the distant mountains. He could hear a dog bark somewhere out of sight. Birds flitted amidst the vegetation. A flight of ducks, honking loudly, passed overhead and landed somewhere down the river.

The only human in sight was tending a distant flock of sheep on the other side of the river.

At the top of the bank he shaded his eyes. The family of aurochs was right where he expected it, in a particularly marshy and unstable loop of the river.

The bull was broadside toward him, apart from the others, a magnificent solitary animal with a gleaming black coat and massive chest and thighs. Mánus had young, sharp eyes, and even at this distance he could see the bull's tail twitch. Flies were plentiful this time of year and could always find tender flesh to bite.

He had been watching this family for some time and knew them by the curve of their horns, their colors, their markings, and habits. A month before, they had been moving slowly upstream from farther down the river. Today there were six cows and five calves as well as the bull. One of the mothers had lost her calf since he had last observed them.

The night he learned Semnali was going to have a child, a spirit had spoken to him in a dream, had whispered into his sleeping ear, had told him what he was to do. The next day he had gone out without telling anyone to begin preparing for the day Semnali would give birth. And now that she was birthing, a power he had never known seized him. He had no choice but to give in to it. He was alone out here, armed with two paltry spears and a rope, and a fierce joy was in him. The mysterious power flowed through, moving his muscles, cycling his breath in and out of his lungs, and absorbing the warm sun into his skin.

Not far from this part of the river was a solitary hackberry tree ten times his height, its massive crown thickly leaved and filled with early fruit. A dense thicket littered with broken branches surrounded the gray warty trunk. Two moons before, when it was still springtime, he had built a small circular corral of goat's thorn there, as sturdy as he could manage.

Now all he had to do …

A ripple passed over his skin. It was not fear that drove him from Ju-Ni and Semnali's labor. No, this was a different fear calling to him! The day had come. Today would not be like teasing bulls in boyhood. It would not be like hunting them in groups, either.

For one thing, today he was alone.

He shook the spears defiantly at the cattle some distance away under an empty sky. The beasts took no notice and continued grazing.

He would do this by himself; once it was done, everything would change. He would be known throughout Ju-Ni.

He knew it would be dangerous to try killing even a calf by himself, especially with that bull standing watch. The bull was a vast beast, the biggest he'd ever seen, and dark, and supremely confident of its power.

But he was not going to kill. He was going to do something more difficult.

Mánus thrilled with tension. The sun was nearing the top of the sky and it had grown hot. He felt like sitting under the tree and sinking into the pure pleasure of setting aside these spears and chewing on some flat bread in splendid solitude.

But Semnali was down in the house with the elders tending her in a world apart from him. Those echoes of old cries from other births propelled him on. What he did today was his gift for the motherbaby.

It was his good fortune that the wind was blowing toward him from the cattle. The power surged in him. He could do anything; he could seize the bull's horns and wrestle him to the ground, single-handed!

Staying low and keeping his eyes on the animals, he made his way toward the hackberry tree from one clump of covering shrubbery to another. He was a wraith, a phantom, a spirit, invisible even in full daylight.

He and the other boys his age had learned this stealth and patience. When they found a herd with a bull on watch, the adults would tell the boys to circle it, capering and yelling and throwing stones and clods of dirt. Sometimes they constructed elaborate headdresses of straw and feathers to look larger and more fearsome, but usually they wore only a small loincloth of leopard skin that flapped behind them when they ran. Once the teasing had maddened the bull into charging, they proved their courage by waiting until the last moment before darting out of the way. He had told Semnali that he was sure the contest was really among the boys themselves and not with the bull.

Since he was old enough Mánus had participated in more than one kill, but he never lost respect for these animals. Even the cows could gore or trample, and he well knew what the bull would do. They were much more than animals: everyone who got near could sense the aura about them. It may be invisible, but it flowed outward from them to shape events, cause misfortune, or confer great joy.

Mánus had picked this massive hackberry for a subtle variation in the texture and color of the grasses following an underground water source from tree to that loop in the river where the cattle were feeding. One of the calves had moved away from the others onto this rich vein. Soon she, for he could see it was a female, was happily cropping the green tops. If, as he hoped, she kept moving at the same slow pace, she would be at the tree before sundown. Otherwise he would have to wait through the night. Either way he was ready. It was coming to pass as his dream spirits had shown him; the calf was coming to him.

He kept low and slow as he had learned, each movement forward followed by motionless waiting. Once the bull coughed when he was in motion and he froze and raised his head enough to see the bull staring at him, eyes even at this distance like obsidian points aimed at his heart. They remained suspended thus, eyes locked, or so it seemed. Did the animal actually see him? Would it charge?

After a moment, the bull tossed his head with a snort and returned to grazing. It had not really seen him. The wind still blew toward Mánus, so he hadn't smelled him either.

Mánus looked up. Blue immensity seemed to flow away from the brilliant torch of the sun. He was in the middle of a plain extending in all directions, utterly alone. The hair stood on his neck. His mouth was dry.

One of the cows looked up and sunlight glistened on the small patch of curly white hair between her horns above the great indifferent blank of her face. He swallowed and continued his painstaking progress, pausing from time to time to check on the bull, and especially the calf.

From the shrubbery under the hackberry tree, he watched the calf. She was small but by no means a weakling. She could kill him if he was careless.

The stream of dark green flowed from a patch east of the tree. Here, just out of sight of the aurochs, he arranged a loop of rope, tied the free end to a thick, low branch, and settled down, the rope loose in his hand. The two spears lay close by even though he shouldn't need them.

Now it was a matter of time for his dream to unfold.

The calf, oblivious, moved closer. It wandered off the stream of grass once and Mánus's breath caught, but soon the calf came back, munching contentedly. Late in the afternoon, she stepped at last into the loop, first one foot, then the other, lips and teeth cropping the grass without pause.

Mánus yanked on the rope, the noose slipped up and tightened on the calf's forelegs, pulling them together. Her hindquarters hopped sideways in a panic. The rope twanged taut and she started bawling.

Mánus had almost forgotten the other animals, but now the cow appeared, running toward her calf, bellowing in rage. Mánus picked up a spear and stood. When the cow saw him, she swerved toward him, horns lowered. He hurled the spear straight and true into her brisket. The butt of the spear dropped against the ground and pushed the point in deeper, snapping the shaft like a twig. She stumbled to a stop, lowered head swaying from side to side, and stared at Mánus with small, resentful eyes. Crimson threads dyed the lines of saliva hanging from her lolling tongue; his spear had penetrated her lung.

In the distance the other cows and calves were thrashing away through the marsh in panic. Only the bull didn't move. He stared toward the hackberry tree and pawed the ground, snorting loudly.

Mánus withdrew into the thicket and waited. The bull trotted halfway to the calf, which, despite her hobbled front legs, had calmed and begun eating again.

The sun neared the western mountains and still, despite the flies, Mánus remained unmoving. The bull went back to the rest of his little herd in the loop of the river, ignoring the silent, wounded cow standing halfway between them, drooling blood. She walked slowly away from both Mánus and the herd. Despite the spear in her lung she would not die right away.

Mánus thought about Semnali and her labors. When he returned to Ju-Ni she and their child would have separated. He pushed this thought aside. It would not do to dwell on the birth: so many babies died to become dimly remembered bones in the foundations of the house.

At dusk the calf would be sleepy and docile, and he would lure her into the corral. For now she was eating contentedly, lurching a step or two with her hobbled front feet.

Shadows were long when Mánus at last eased himself out of concealment and untied the rope. The calf looked up at him with large, blank eyes and he stopped. She blew air through her nostrils and returned to the grass.

He approached her, coiling the rope. Whenever the calf looked up, he stopped. When he was close enough to touch, he reached out slowly. She jerked her hindquarters away. He stopped.

When the flies were particularly vicious just before dark, he brushed them gently from her back. The light winked out, and a full moon burst above the distant mountains, washing the sea of grass with silver. The cow had wandered out of sight or sunk at last into death. At any rate he could no longer see her. He would have to remember to send someone from Ju-Ni to find the body. She could feed many people.

He looped the free end of the rope around the calf's neck and coaxed her step by step to the corral. Progress was slow, with many stops, but he finally closed her inside with plenty of grass.

His dream spirits had spoken true: he had taken the first step to tame an aurochs, the most fearsome of animals. He would lead her back to Ju-Ni and when she was old enough, he would stake her near the wild and let her mate. He would then raise her calf. She would join the sheep and goats at the settlement. There would be no need for dangerous hunts any more.

Satisfied, he lay down beside the corral and went to sleep.

A low growl woke him. His eyes snapped open to blinding white light. The vast bulk of the bull swam into view a few steps away, head lowered, polished horns curved toward him. His hot breath surged loudly in and out, close enough to blow damp warmth onto Mánus, sprawled against the corral's fence of saplings and goat's thorn.

Morning mist rose around him from the dew-drenched grass. The sun was a blinding smear above. He struggled to sit up, reaching at the same time for his one remaining spear, a paltry weapon against this massive beast before him.

The bull rumbled low in its throat. Once long ago Mánus had heard a leopard make a warning sound like this.

The other cows and calves were gray, insubstantial, nearly transparent shapes in the mist. Only the large blank pitiless eyes watching him seemed real. Mánus tore his eyes away from them. He tried to avoid looking at the vast bulk menacing his world, black as the bear outlines on the small blanket, that moment he had seen the dark showing through.

His breath came shallow and quick and his heart thudded rapidly. The sun raced across the sky; it would soon be dark again and he would be safe. But no, he was wrong, it was a motionless blur above the two distant peaks at the eastern horizon. This day was just beginning. He released his breath slowly and reached for the remaining spear, thinking of the broken one. He could still hear the cow's breath and the small sound of it snapping (where was she now?). Could he dodge around the hackberry tree and evade the bull? How long could he lie motionless like this, fear frozen in his throat? He had to move, escape or attack.

The calf suddenly butted against the fence behind him, bawling loudly.

Mánus started to his feet.

The great aurochs lurched toward him.

As he struggled to his feet Mánus understood the bull was intent on breaking down the fence and freeing the calf, and that he was simply in the way. He looked at the bear Semnali's foremother had pressed onto the back of his hand. The sinuous black lines opened to a darker world, for it was the last thing he saw before the bull shattered his chest with a long-hooked horn, cracking the sternum and ribs, knocking him aside with his massive shoulder.

Mánus died instantly.

Three days later some women from Ju-Ni were foraging on the plain when they found his body and the spear. Since they did not know who he was, they buried him in a shallow grave inside a circle of flattened goat's thorn. One of them suggested it was a corral, though why anyone would build such a thing in this place none could say.

When they returned to Ju-Ni with the spear, they learned who he had been.

Only months later did someone find the remains of the dead cow. There was very little of her left, but they carried her head and horns back to the place with them.

They said nothing to Semnali's foremother and the others about the possible corral or the dead man's crushed chest. The corral was an unsolvable puzzle; Semnali and the unborn child were already reborn in the underfloor; the Song of Farewell had seeped into the walls and was gone. What use to speak of such things?

Most animals amenable to domestication had already been drawn into the human world, enclosed in walls and dependent on human nurturing. Many can no longer survive without it.

Cattle and horses remained untamed. At this time, the aurochs, both a fearsome natural force and a potential bonanza of protein, topped the list of animals ripe for domestication. Though less docile and herd-bound than sheep and goats, they offered such a richness of resources from meat and milk to burden-bearing and leather that they could not be allowed to remain wild.

Once domesticated, thousands of years of selective breeding have made them docile, nutritious, and delicious. They have also rendered them high in fat, the way grains have given us too much starch, and inadequate to survive on their own. We are locked in mutual dependency.

Mánus may have failed, but he was not alone. Across the region attempts were succeeding, and soon cattle had joined sheep, goats, and pigs in the agrarian compound. From then on humans would determine their fates. Horses, donkeys, camels, llamas, turkeys, pigeons, chickens, guinea pigs, silkworms, and many others would fall into step.

The Beast Within

Once tamed, the aurochs would no longer belong to the external world. Brought inside and folded into the life of the household, it became another domesticated resident.

Sometimes they were painted, a way to keep alive the memory of the mighty aurochs's violent interactions with settled humans.

After the deaths, Semnali's foremother kept the small blanket neatly folded in her own house. The next house Edgeward, Semnali's, remained empty. People whispered that it held too many unhappy spirits. Over the following months and into the cold season, when hard, dry snow rattled against the leather lean-to that protected the entrance from the wind, she thought about Semnali the motherbaby, and Mánus the father. She knew of some who might in time want to occupy the house, but for now it was a hollow shell, a husk bereft of even the feeblest flame of life.

Many a long afternoon she took the blanket, climbed down into the house and built a small fire in the oven. She sat on the platform where she and the others had buried the motherbaby and ran her misshapen fingers over the marching bears on the blanket's folded edge, trying to coax life back into the emptiness. She knew it was there in the bones underground, she could feel it. But others felt something stronger: a vague, uneasy sense of alarm. Unquiet. Something was missing, something left undone.

Spring brought wet, and then dry. Even when the three still-living elders sat with her, she seldom listened to their words or answered their questions. Then one afternoon during the hot season when the year had come around again, she broke a long silence. "It is time." Her fingers, tracing one bear after another along the blanket spread on her lap, stopped at the last.

The surprise of the elders soon passed; they understood she meant Mánus.

"Time for the house to live again," they agreed.

They climbed to the roof.

"Tomorrow!" Foremother pointed at the sun casting long orange shadows over the roofscape toward the hackberry tree and Mánus's shallow grave. "We must bring him home."

It was as she said. At dawn one of the foragers who had helped bury the body took Mánus's younger sibling, Little Brother, cheek shaded with his first beard, out to the hackberry tree. The corral had disappeared, and the grave had subsided into the ground, but the older man knew where they had placed Mánus on his side, knees drawn up.

His flesh was nearly gone, but the long bones still held together. The man showed the boy how to sever the skull from the spine and the limbs from the shoulders and pelvis. They bundled up the limbs and skull, leaving the spine with its shattered ribs. Mánus had no further use for them.

Under Foremother's direction, Little Brother dug into the sleeping platform and uncovered the motherbaby's head and upper body. As with Mánus, her smooth skin and dark eyes were gone, as well as her swollen belly. Some fragments of her lustrous dark hair were all that remained around her skull.

"Bring up her head," Foremother commanded.

Little Brother had first handled bones that day, and in his clumsiness he scattered the first few cervical vertebrae. He apologized and handed the skull to Foremother, who smiled to reassure him.

"This," she intoned, holding Semnali's skull in her hands, "is the life of the house. We bring it into the light. We bring Semnali into the light, into the air, into the house." She gazed into the empty eye sockets and inhaled deeply, taking in the sweet scents of soil and decay. "Her spirit, their spirit, has flowed into the bones of this house. Here it—they—will remain. Put Mánus's bones on her body, boy; lay his skull in her head's hollow. After we cover them, they will become spirit-in-house, three-in-one, and the house will live once more."

The boy, clumsy still, let Mánus's skull slip facing backward into the hollow left by Semnali's. He quickly arranged the long bones flexed over Semnali and filled the new grave. His hands were trembling. When he had finished, he sat back against the wall, struggling to control his breathing.

Foremother touched his shoulder and stood. When they emerged onto the roof she gestured to the chosen couple and their two young children. "Plaster the walls again," she commanded. "Move in your things." She handed them the motherbaby's skull. "This is Semnali. Plaster this skull, give her a face, and she will watch over you. She will watch your children, she will make them strong, she will send you more. Her power flows through the house. Go."

Once the small family had disappeared into the gloom below her feet, Foremother stood in deep thought while the others waited. "Tomorrow we go to the house of Ueik of the Weeping Trees," she said at last. "Today Little Brother will polish the walls of his house to make them ready for Ueik to manifest the story of the Bull, and Mánus, Semnali, and the almost-person."

The next morning all clambered across the roofs, Foremother, the elders, and Mánus's younger brother, to the house of Ueik of the Weeping Trees, which was on the eastern side of Ju-Ni away from the River, low down on the slope near the Edge. Beyond was the Open where large parties assembled to feast at the solstice or after a particularly good hunt. Most of the Weeping Trees had houses nearby.

Ueik was a round, happy man with red-stained hands. People often called upon him to paint on their walls, usually to acknowledge a birth, a death, a great feat of hunting. This was why he had no need to grow crops, hunt, or forage for food. When he accepted, people gathered on the roofs nearby to discuss what he might paint down there in the gloom of a house. Would he paint patterns, handprints, or other stories? He never created the same thing twice.

He often left Ju-Ni for days, even moons at a time, to return with lumps of soft red stone to grind into paint. His other color was charcoal black. Down in a house he would mix his paints in two precious shells carried all the way from the bitter water in the south. They still smelled of salt. He made brushes from the frayed ends of certain willow sticks or, when he could get it, the bristles of a wild boar.

"In Mánus and Little Brother's house," Foremother told him. "A story."

"I know the house," Ueik answered with his usual happy grin.

"Yes," Foremother agreed. "You know the house, the story that binds them."

"Of course." He retrieved his basket of paints and brushes and again the small troop climbed up and down over the irregular roofscape to the house where Mánus's younger brother still lived with the old woman, who had birthed four live children, two of them now dead. She could barely speak, barely move, but she grinned and nodded when they come down the ladder and Little Brother put his hand on her shoulder.

Ueik looked at the eastern wall. It was white with fresh plaster and reflected the sunlight pouring in through the roof hole. He squinted, the corners of his mouth twisted up, first one side, then the other. "There," he said, pointing at the lower part. "Deep under the plaster there is a stag, his tongue hanging out to one side, exhausted, dying."

"This time," Foremother said, "it is an aurochs bull, the biggest ever seen, the one who killed Mánus. He was a brave man, Mánus, companion and father of our motherbaby. He faced the bull with a single spear. The bull killed him, and he has become the bull spirit, the two of them, man and bull, blended as semolina stew is blended with meat and fat, and welcomed below by the headless mother."

Ueik nodded while she spoke, murmuring, "I see, I see," when she paused. Now he waved her away and sat against the wall on the opposite platform, looking at the empty white expanse.

"There is a handprint over there." He pointed to the northeast corner.

"That was before the stag, and you still remember?" she asked.

"I was young then. I watched my uncle ..." He unpacked his tools and began mixing, already in his trance. Foremother waved the others up into the bright summer light. She, as eldest foremother, had the right to remain behind and observe the artist.

Little Brother had done well; the plaster's whiteness dazzled. It seemed to Foremother as if the light came forth from within the wall. And perhaps it did.

Ueik drew with the tip of his finger near the top of the wall. "Here," he murmured. "Long, long ago, at the time of the stag." He looked up at Foremother. "A line of men." He traced the fingertip along the wall, pointing at one invisible figure after another. "Still there under the plaster, of course, dancing, hunting. Always were, always will be." He sat back on his heels and held up his brush. "Speak. Tell me."

Foremother could see a line of identical bears marching along the edge of a tiny blanket. They too were hunters, dancers. They invoked the aid of invisible beings. She closed her eyes and took in a long breath. Something like smoke swirled. She had not lit the oven today, so it was not in the room but inside her lids. "I see the bull," she said softly. "Pawing the ground. He is angry."

"Of course," Ueik murmured, dipping the brush into red ocher paint. "The hunters are teasing him. They were boys become men. They were Mánus."

Below the line of hunters painted over by countless layers of plaster and now recalled, he sketched the aurochs. Or, Foremother believed, his brush uncovered the beast; he flowed wet onto the plaster, facing to the right. His drying bulk occupied much of the bottom half of the wall. He grew larger, more powerful and belligerent. His thick legs braced as if someone were trying to pull him forward and he was resisting. The brush flew back and forth to the shell of red, filling in the enormous body, a tail with a brush of hair on the tip, two fierce curving horns, a lolling tongue. The bull's penis was aggressively erect. He was fertility and life as well as death.

Foremother heard the animal's breath and felt its terrifying heat. She opened her eyes. It was as she had seen. She closed them again.

"There was a corral," she murmured. "And a hackberry tree."

The brush remained still. One did not paint mere plants, only wild animals and active men, to express how they related to one another. Anyone who knew the story would see the hackberry tree, the corral, the river curving far away, a family of animals watching.

The old woman murmured, "Semnali is there."

Ueik moved the brush and a headless woman with a swollen belly and heavy breasts appeared below the bull. Foremother's eyes were still closed, and she did not see this woman until she opened them again. She had said nothing to Ueik about taking Semnali's skull, yet he knew, and understood, for that woman near the mighty bull was surely Semnali as a vital and gravid motherbaby on the verge of her labors.

Ueik took a new brush, dipped it into his shell of black pigment, and manifested the small figure of a man holding a spear, spotted leopard skin loincloth splayed out behind him as he ran toward the bull. He was smaller than the woman, puny against the great beast.

Foremother reached out and took his upper arm. "He was alone," she said. "He and the bull exchanged spirits. He became the bull. The bull became him. See how you brought him forth. He is not throwing the spear. You have been there, Ueik, and done well."

"The hunters above, the dancers below," he began, but she squeezed his arm and he stopped.

"There is a bend in the river near the hackberry tree," she said quietly. "A place of aurochs. Mánus knew that place." She nodded her gray head. "He spoke with the animal that killed him." She looked at the three figures on the wall, the bull, the headless woman about to give birth, the hunter. "They tell us things, all of us. They speak of great changes." She blinked and shook her head to clear it. "Great changes."

Ueik looked at her. He could not tell if she spoke with hope or regret.

The custom of removing skulls for special treatment was well known long before. After the flesh decayed, the living retrieved the skull. Sometimes they covered it with fresh plaster and repainted it as a social form of veneration for a lineage ancestor or a private obsession with a beloved dead, as in a story by Edgar Allan Poe.

A plastered and repainted skull like Semnali's would become the tangible, reanimated presence of someone long gone, a precious, even sacred representation of an episode in the biography of the house embedded into the psychic lives of those still alive. Sometimes skulls were (re)buried in a house foundation or under a support post, transforming the dead person into a true founder.

A mural similar to this one survives. Many people actively tease an enormous bovine. A story unfolds before the eye.

As in the Franco-Cantabrian cave paintings (and unlike the more public low-relief animals at Göbekli Tepe), the small, crude stick figures that appear often on the house walls at Çatalhöyük were not meant for large audiences. Few people could fit at one time into these dark, semi-subterranean cells. In many of the paintings that remain, men are running or dancing, waving their bows or pulling the wild animals' tongues or tails. The figures take many positions, sometime on all fours, or standing and pointing, capering, or gesturing, or high-stepping like a chorus line. Many wear leopard-skin loincloths that fly out to the sides behind them. They are reminiscent of the loincloths worn by the enormous man-pillars at Göbekli Tepe.

Single women loom larger than the men in some of these depictions. Though present they do not take part in the men's activities. The headless woman standing beneath (or in front of) the bull in one of the most famous paintings is clearly pregnant. We can connect fecundity and abundance with control over the world outside. The nonhuman world has moved outside the house walls for good.

Of Pots and Plans

There came, as always, a time when community life unraveled. Perhaps it was a change in climate, for a period of cold had come upon them. Other towns in more favored places were growing in importance. Perhaps new zoonotic epidemics decimated the populace, for more and more diseases were spreading from closely held domesticated animals. Perhaps some important local resources were exhausted.

Whatever the cause, population fell. Atop the mound, houses moved farther apart. Space opened around them. They had doors.

Pottery—mud transformed by fire—took up the duties of other, less amenable materials like leather and wood.

No new technology, though, could save Ju-Ni.

Two—the man Legh, branch of Waterbird by way of Little Brother, and his woman Oner, last Dream of Twin Mountain from a cousin of Semnali—stood together on a terrace of their house.

Down slope, beyond the last dwelling a light blanket of dry, meager snow on the fields and frozen marsh was taking on a burnt orange tint like drying blood. A few winter crows circled overhead, cawing uncertainly. They settled only to rise and circle again before at last settling into uneasy silence on a jagged wall.

It was just past the height of day and far from dusk, yet the light was dimming. A dark curve had cut into the disk of the sun, and people from the surrounding houses, people of rival clans like Scorpion and Deer, were whispering that Legh and Oner, last of their clans, had offended Great Vulture, so Great Vulture was swallowing the sun in retribution. Their whispers were loud and angry, intended for the old couple to hear.

Oner took Legh's hand. Be strong, her grip told him. Ignore the whispers. We know what we must do.

They were old, the two of them, older than most around them, older than they should have been. Their lives had been long. Yet, had anyone asked, they would have said their ancestors had abandoned them, for they were the last of their clans.

Smoke rose from still-living houses, but many gaped, empty and open to the sky. Dust had sifted into them for years and lay banked against the walls. Dry snow filled the cold, rectangular hearths. Their doorways were mouths distorted in death.

Despite the ample grain and hackberries in their own storage bins, the dried meat hanging from the roof beams, and the jugs of water brought up from beneath the river ice, Legh and Oner felt no hope. A bone-chill had fallen over Ju-Ni half a lifetime ago and not once had it loosened its grip. Rains teased but seldom fell; crops withered in the fields. Every night for three moons last summer a pale streak hovered in the sky and cast a deathly pallor over the marsh. People saw its image doubled in the river and told one another the year would be a bad one again. In the fall a sickness had burned through the flocks, killing the greater part. Several people had stumbled and fallen from roofs or retreated to their beds, never to rise again. Some said a two-headed aurochs calf roamed near the ancient hackberry tree where a bull had given the gift of supernatural power to Mánus of the Waterbird clan, who had been a man and became something more. For generations after, Waterbird was strong and the name of Mánus, like Semnali's skull, renewed the vitality of the clan.

Now, though, Legh and Oner's own childless children were dead, and the long lines of their clans would disappear: all those houses, all those generations, gone forever.

And Great Vulture was devouring the sun itself ...

Generations of old women had told of Great Vulture tearing at the sun and pulling away pieces of its flesh, but everyone knew these were just stories to frighten children. For as long as Ju-Ni had been, it was said, the People had offered the heads or bodies of some of their dead to Great Vulture, and always Great Vulture had circled and settled, had cleaned the bones of the dead and been satisfied. No one ever believed Great Vulture would swallow the sun whole.

Legh knew they both felt the same dread. They had sung their clans, Waterbird and Twin Mountain, down into the earth. They owed it to the dead to do this publicly for the first and only time. They had to expose to everyone their clan rituals, the very core and meaning of clan life.

This was bitter. It was doubly bitter to be surrounded by whispers of their failure and shame as they gave away everything sacred, as was appropriate when a clan came to an end. In this instance, two clans. Though they did it for the ancestors, it hurt.

They both would rather die, though they did not say this aloud. They would have given much to go under the earth and become bones of the next house, but this was impossible. They must live. Clan and lineage spirits demanded they finish what had begun so long ago.

Legh stirred. "There is the lamb," he murmured.

Oner hissed softly over her arched tongue. Of course there was still the lamb, the last of the flock since the demon sickness came.

They went down to the courtyard. Oner disappeared inside the house to bring out the precious four-faced pot and ancient plastered skull.

Legh had lifted the lamb from its pen and now stood, chest heaving with the exertion. He set it down and watched the animal wander to the edge of the court. It looked down into the garbage pit, bleating pitifully. Legh gagged back the sour taste of failure.

Oner set the skull on a low wall facing the house, took a knife from the pot, and handed it to Legh. "Be quick." She was gruff, and looked away.

The obsidian blade, a rivulet of darkness, flowed from the bone handle. He held it up and a blood-red highlight ran down the cutting edge.

Oner cupped her hands over the bull's heads molded onto the rim of the pot. She waited in stolid silence until her patience gave out. "Get on with it," she snapped. One of the human faces on the pot looked sadly at Legh. The other was pressed against Oner's belly.

All that remained of the sun was a shimmering curve of orange light on the left side, shaped like the horns of a great bull. The crescent winked out and the sun was gone. A crackling halo flared around the darkness. One of the bystanders breathed, "It's Great Vulture's head!"

The couple wanted an end to their bitterness and anger, but Legh hesitated. Other clans like Scorpion and Deer survived. They had children. They had a *future*. Twin Mountain and Waterbird, Legh and Oner, had none. The cycle of their lives had closed.

The lamb, its shabby wool dyed an uneasy orange by the strange light, looked at Legh with bleared, rheumy eyes when he yanked up its head. Pressing his lips together, he struck at the throat and the lamb went down with a sound no louder than a child's sigh.

Oner caught blood until the pot was nearly full. The rest of the animal's life soaked black into the dirt of the court.

She held the pot up to the flaring circle of fire and spread her fingers so the bull could see Great Vulture's dark head. Slowly she turned the pot so the bulls and the two people gazed at the circle one by one. A hush fell over the small gathering. All, men and women, were staring up, mouths open.

The sun had stuck in Great Vulture's craw. The bird glared, flares crackling around the shadow of its smooth, round head. The pot began to shake, to surge with pops and bubbles. Steam writhed above the roiling lamb's blood. The mournful faces on the pot, human and bovine, looked down at the earth. They were molded clay and said nothing.

"It's the end," Legh intoned with a shiver, his voice nearly inaudible.

Oner touched the lamb's blood and jerked her finger away. Its tip burned red like a flame. She placed the tip against Legh's forehead and his flesh sizzled, leaving a livid mark in the midday twilight. He flinched, pressing his lips together. She tapped his cheeks, first one, then the other, and tendrils of smoke seemed to rise from the small, round marks, like burns.

She held the pot toward him so he in turn could touch the blood. He jerked away; his fingertip had become the wick of an animal fat lamp sending up blue flame and smoke. The people, insubstantial as shadows, drew closer. He felt their envy and hatred and held up his burning finger. They breathed out one long, collective sigh and backed away.

Just as Oner had touched his forehead and cheeks, so Legh touched the wooden lintel and jambs of the door. Suddenly, the hearth fire, deep in the inside darkness flickered to life. The old couple thought they heard a low moan that rose, shrieking, and burst into a cascade fragmenting into purple and blue lightning.

The others witnessing said nothing. Legh thought they must have seen nothing.

Oner, bowl crooked in her arm, then handed him the plastered and painted skull.

The moan subsided to a steady drone, and against this Legh began to chant, quietly at first. When the circle of fire around Great Vulture snapped out and a new crescent appeared on the right side, his voice rose. Light seeped back into a world utterly changed.

He sang for the last time the founding story of his clan. He told how Semmi had descended into smoke and terror and how Twin Mountain had received her. He told of Lokke and Lokkani's great trek to the place that would become Ju-Ni. Lifting the skull, he told of motherbaby, saying this head in his hands had belonged to Semnali of the Twin Mountain clan, and that she had offered her lives, old and young, to the clan's great song.

When the story shifted to Mánus, Legh handed the skull back to Oner.

The others, the men in thick fur, the women wrapped in dark wool cloaks, fell silent, caught in Mánus's story as if it were their own. They heard the bull roar its rage, felt the thud of its great hooves on the thick earth. They cried out when the cruel horn smashed through Mánus's chest and took the breath of life from him.

Legh and Oner always told others that the human faces on the pot were Mánus and Semnali. The bulls united them in a circle of life, brow to horn and ear to ear around the bowl of it. Mánus became Bull in the moment of his death, and Bull, indeed, all bulls that had given their lives to feed the people of Ju-Ni, carried Mánus's powerful essence down through the years, and through the seasons of the years, to this day when Great Vulture swallowed the sun and vomited it up again until it shone down on a people transfigured.

"It is an end," Legh said. His shoulders sagged.

Oner shouted, "An end," and threw the lamb's blood over the threshold into the house. It hissed and bubbled in the darkness, sending out threads of sooty smoke that drifted silently into a clear, ice blue sky.

The hearth fire inside had spread to the walls, and now with an agonized wrenching of wood, like a squeal of pain, a section of roof collapsed. Slow, billowing clouds of dust spread across the court. They blew outward with a deafening bellow, pursued by orange flame that licked lasciviously at the lintel. The roof erupted in gouts of flame. Ash and debris swirled around them.

"The end is the beginning," Legh shouted. The words were dust in his mouth, for he knew here was no beginning.

The others had backed away in horror.

Legh nodded at Oner, who tossed the four-headed pot into the garbage pit, where it shattered.

She tucked Semnali's skull under her arm and together the old couple walked through the crowd and descended the slope through the other houses, those still standing and those that were ruins. They crossed the frozen river, and so carried the skull away from their burning house in the dying settlement of Ju-Ni.

Great Vulture was gone, and soon they, too, were lost in the snow-dusted waste of dried shrubs and frozen marsh on the other side of the river.

"They will die out there," a woman remarked.

"Yes," answered another, with satisfaction. "They will die."

When a settlement ends, it takes its stories with it. The shrinking settlement of Ju-Ni now huddled atop the mound amid abandoned houses and trash pits. When Semnali's skull moved on, it took with it all the memories that make up the history of her line.

Ju-Ni would not survive much longer as the practice of skull preservation was losing its utility. Other forms of memory management would take over. A couple of millennia later a remarkable innovation for storing memory externally would arise some distance away in what is now southern Iraq.

Clay grew ever more ubiquitous in bricks or crude animal and human figurines. Learning how to fire clay containers dramatically improved cooking and storage. A ceramic pot could be left on the fire; the stew no longer needed constant attention. Yet, this new innovation helped kill Ju-Ni.

Fragments of the extraordinary four-headed pot were found in a trash pit at Çatalhöyük. Once reassembled in the laboratory, the faces, two human and two bovine, look out in the four directions. Their ears are shared, and the animals' horns curve around to become eyebrows, though the plump human faces are eyeless. They have sharp noses and small, puckered mouths, mimicking the plastered skulls. Cattle, safely domesticated, were no longer fearsome adversaries. It was a domesticated human who shaped them into an ordinary pot.

III

HOME

An Introduction to Origins

We've arrived at the threshold of modern life and the origins of history. Pottery led to metal working, which ushered in the Bronze Age. Settlements grew into cities and houses became homes, the center of daily life, surrounded by social and symbolic spaces of many kinds. Cities included public buildings, communal spaces, complex hierarchies, and extended families. Nomads, bandits, and outcasts were relegated to the wild steppe outside the range of the city's growing influence.

Those who tilled the lands nearby could no longer simply move elsewhere when conditions turned against them. The wild rapidly disappeared under the plow or the woodcutter's axe. Increasingly, farmers were trapped by their investments in preparing, plowing, and irrigating the land, building permanent houses, managing domesticated herds, not to mention all their increasingly complex economic ties to the city. They had lost the ability to survive without all their entanglements.

Social complexity, inequality—all the ills that man is heir to—forced humanity to look back. How could they explain their origins and all the steps that brought them to this place?

There are two main answers. They come in the form of origin myths.

One, carried out of Africa as much as sixty thousand years ago, lingers mainly in Australia and Southeast Asia. It has little interest in what happened before people: they erupted out of the earth or broke out of an egg. Nothing came before. Its concerns are with the relationship between human groups and the territory they roam. It stands outside of time, following the connections forged over thousands of years with its animals and places. In Australia it was called the Dreamtime, without beginning or end.

The second myth arose more recently in Europe and Asia. From Iceland to Tierra del Fuego, from Spain to Japan, bards have spun variations of the world's beginning from primordial chaos through generations of gods culminating in mankind. People, through some fault or crime, were nearly destroyed, often by flood. This should not be surprising in a sedentary world dependent on reliable water.

This myth is an allegory of a human lifetime: born in moist darkness, growing into knowledge, learning by making mistakes, and suffering through disease, old age, and death. Just as the end of an individual life is inevitable, so for the world. Even today millions believe the end is near.

The Neolithic, like all such transitions, was slow, uneven, unexpected, and largely unnoticed. Towns grew into cities, which were crowded, noisy, unfair places of squalor and cruelty, dangerous in ways the predators and natural disasters of the long years of the Paleolithic were not. Of all mankind's enemies, mankind took first place.

City dangers are human: deforestation and destruction of the environment, disease that jumped from domestic animals to humans, inequalities inspiring new forms of interpersonal violence, more elaborate social stratification, and systems of oppression sustained by an emergent class of religious experts who codified the myth into an instrument of politics. Then of course the predation of cities on cities and the emergence of empires.

Gods gave humans things they needed. They (usually a male) brought fire, showed them how to catch fish, and how to build a house. Sometimes a goddess appeared with gifts, weaving, for example, or sexual pleasure. At the same time, the gods punished, because suffering exists.

Heroes, who could help the gods keep mankind under their control, took the lead in explaining why the world needed to be just as it was. Doing so became a calling, which they passed on to their children. In this way, the world will continue until its destruction by fire.

Until recently, no one noticed the inevitable, painfully slow destruction of the world itself. Levels of CO_2 and methane began to rise unnoticed with the spread of agriculture. The diversity of species gradually yields to monoculture. The land in Mesopotamia sickened under the burden of salts leached from below by plow and irrigation canal. Plagues and crop failures grew increasingly common. Species and resources dwindled; some vanished.

Welcome to civilization.

The Need for Walls

A thriving culture in the resource rich marshes at the confluence of the Euphrates and the Tigris Rivers gradually spread northward along the waterways flowing through the desert. The rivers provided clay for ceramics, water for irrigation, and convenient transportation for the long-distance exchange of wood, metal, and stone. Silt from seasonal flooding made for rich soil, which needed irrigation. Ditches became canals, seeds demanded plows, and plows were useless without domesticated animals to pull them and people to direct the animals. All these needed organization.

Once a village became the focus for its region and developed neighborhoods of increasingly rigid divisions of labor with specialized buildings, it became a city.

Cities were juicy targets for human predation and soon became candidates for defensive walls. Another myth had to emerge, a local myth, of the man who built the walls.

When he was a child, Merkar's father, Meshkiangasher, moved his family into town to be closer to the Kulab. Unug was a small place then, just a score of houses, maybe one or two more. The largest belonged to the sky god, An. On festival days everyone would pay respects and feast. Other times, when the barley wouldn't grow straight or Enlil held back the rains, Merkar's father went alone to speak to the god. After all, An was Enlil's father and should have interceded with his son when asked.

When that didn't happen, Merkar's father cheerfully announced that the gods were capricious and willful, that was their way. They wouldn't do everything for men, he said. For the grain to grow men had to bring the sweet water to it with the sweat of their labor. So every morning Merkar and his father walked down the river a double-hour to the fields, stepping over the small watercourses and wading the wider ones. They worked through the day and walked back in the evening.

One warm, calm day after Utu the Sun had gone below, the bright star of his sister Inanna glimmered out of the darkness and appeared reflected on the smooth surface of the canal. "I've been thinking," Merkar's father said. "If we really want help with the barley, well, for that we should call upon Enki, who taught farming and herding to the people. But Enki has no House nearby where we can make the proper offerings."

"Why don't we ask Nammu, who gave life to An and all the others?" Merkar asked. "Or Ninmah, the Mother of All Life? Why don't we ask them?"

His father's face darkened. "Because they're busy and have no time to hear us complain! You are benumbed by the gods, boy." He waved at the firelight in the windows of the houses up ahead. "An is *here*, in Kulab, in his House, so we ask *him*."

Merkar never questioned his father again. Like parents, the gods had created men, and like parents the gods owned them. As he knew from his father's example, parents don't listen to children, so why should the gods listen to men, who were no better than the slave women from the mountains working at the looms!

It was a puzzle, especially when, despite his grumbling, Merkar's father continued to take the family to An's house every ten-day. Leaving them outside, he took something into the long, square structure, and came out empty-handed. One day it might be a rabbit, another, a *sila* of barley or a jar of palm oil.

Many times he told the family—Merkar, his mother, brothers, and sister—that he was a trader as well as a farmer. He confessed that even more than a good harvest, he asked An for guidance and safety on his journeys and help in his dealings with distant strangers.

Often, he was gone for months at a time, leaving the rest of the family to plant and irrigate, weed and harvest. Sometimes he returned with baskets of small colored stones or flat ingots of copper; other times he came back with an empty boat. "Still, I'm a better trader than farmer," he repeated cheerfully. "The gods favor me."

Then came a time when he went south through the marshes to the bitter waters of the salt sea, a journey he had made many times. They watched him disappear into the thick stands of reeds, waving cheerfully, but this time two rounds of seasons went by, and a third, and still he did not return.

By then Merkar had become a man and a more capable farmer than his father. He had himself purchased a small statue of Enki and spoken to him even though the god had no House nearby. He had asked what to do, and Enki, protector and benefactor of mankind, inspired him to build catch basins and sluice gates to control the water. His harvests increased, and his family lands had grown. When a place grew salty after a few years and nothing would grow, he channeled the water in a different direction.

Because their labors were so trying, Merkar understood why his father preferred to be a trader, to see other lands, to speak other tongues, and to bring back things people wanted.

Then one especially good year two moons after the New Year festival, when the harvest was in and the land was dry, he loaded two talents of barley from his storehouse on his donkey and led it through the maze of streets and canals of the town to the House of An.

Unug had grown much larger by then, with many houses.

He led his donkey up the broad staircase of precious limestone to the House at the top. From there he could look out over all the city and the broad river to the southwest.

The priest was waiting by the wooden double doors. The doors of precious cedar were flung wide, revealing a tall rectangle of darkness set into the brilliant white walls.

Merkar tapped his two great baskets of grain. "For An," he said.

"You are generous indeed, Merkar." The old priest signaled for two young men to unload the donkey. When the priest grew too old to farm, he had dedicated himself, along with his two slaves, to keeping the god's house.

"For us, An is the most important of the gods, is this not true?" Merkar was remembering his father.

Utu the Sun poured light on the polished lime plaster of the walls. Merkar shaded his eyes and gazed across an empty field at the E-anna, the House of Heaven a few hundred paces away. It belonged to Utu's sister, Inanna.

"Of course," the priest agreed. The two young slaves hefted the baskets of barley onto their shoulders and disappeared inside. "Come," he said. "You can thank the god yourself for your harvest."

Merkar hesitated.

"What is it?"

"You see the E-anna, the House of Heaven, over there," Merkar mused without moving. "Inanna, too, is important. Sometimes, and many people have told me this, she seems more responsive to their pleas than the … other gods. Perhaps, they say, this is because she is the daughter of Nanna the Moon and sister of Utu the Sun, who are both closely involved in our lives."

"Perhaps." The priest tried to conceal a growing sense of alarm at what might be a slight to his patron god.

Merkar noticed and said soothingly, "Of course, to survive we need all the gods. An has been the father looking over our city since the beginning of time. The Kulab is his home, as Unug is my home also, and yours. Remember how it was when I was just a boy and you were still harvesting dates and urging barley to push out of the dirt each winter? There were few houses here then. This was a village like all the others up and down the river. Now we hear metal workers' hammers through the day and smell the smoke of their furnaces. An has much to do these days."

"And?"

"And over there, all that remains between the two towns, the town of the Kulab and the town of the E-Anna, is that field. In a cycle or two of seasons, houses will fill it and the two Houses of the gods, the Kulab of An and the House of Heaven where Inanna lives, will be separated forever."

When the priest bowed his head his scattered scalp hairs bent like barley stalks. "What do you suggest?"

Merkar grunted. "Kulab and the city Unug are one. Unug will grow that way, past Inanna's House. Can we prosper with two great Houses that would divide the city?"

"What?" the priest cried. "You cannot think ... ."

"I do not say the two Houses should become one, certainly not. What I do say is that if we are to be one town, then where we are standing, and what we see over there, and the field between them, must become a district for the gods alone. An and Inanna must have their place, as the potters and the metal workers have theirs. This field between should belong to both our gods, a place for all to gather before whichever god they choose."

"An has no need for fields. He has much land outside of town," the priest began. "He does not ... ."

"Yes, yes, An has his lands, as Inanna has hers. But we bring gifts to the gods here because we want to speak with them directly, is that not so? Inside these doors, as I know from many visits, the god waits to accept my gift. It is the same with the E-anna, where people speak with the goddess. If these two places could share this open space ... ."

Merkar chewed on his lip for some time. "There are men who have little," he continued. "Men who would take what belongs to the god to feed themselves, to feed their families. You have seen this, I think, in your long life."

"Yes," the priest murmured. "More often than I care to say. It does not go well with such people if we catch them, though."

"The houses of the gods should be protected, don't you think? Good houses leave nothing open for others to take. Is it not said, 'The owner of a house should reinforce the windows against burglars?' Those who attend to a god's house should do no less, don't you agree?"

"You speak of walls? Like the one around your house three streets south?"

Merkar smiled. "My house, yes, where I have lived since I was a child. We all have seen Kulab and Unug grow thick with people the way barley drops seed and thickens, the way a family's house adds to itself year after year, and year after year we must expand the walls. Tell me, how far do people from the countryside walk to bring their offerings for the Kulab?"

"I'm told some walk as much as half a day."

"Half a day! And half a day back home again! Many of us who live in town walk out to our fields as far to tend them. Now tell me, besides their devotion to An, why do people do it? Why do people come such a long way to give to the god?"

The priest pulled thoughtfully on his lower lip. "It isn't just to see the god. They find things they need here. They trade barley and dates for metal ornaments, for pots and plates."

"Yes, yes, but they could find such things in their own villages," Merkar replied. "Here the metal workers live in the street of metal workers, and those who make the pots and plates live near the others who do the same. More and more, men no longer have lands to farm, yet they cannot eat their pots or drink their tools. So they live on the barley others bring them."

"True," the priest agreed.

"Now you can see that Unug is becoming Kulab *and* the E-anna?"

"I see."

"And do you see that the Kulab and the E-anna are rich with offerings?"

"Yes."

"There are people between the rivers who might find such riches too great a temptation to resist. They might be our own people or from other cities. This is the reason the House of Merkar is surrounded by walls."

The old priest gave an anxious look around. "Come inside."

They passed through a small entrance chamber into a long, central room. The brick walls and the ceiling of tightly packed reed supported by orange-painted wooden rafters were lit by a series of torches along both sides. At the far end a large platter of crimson fire burned before the statue of An, whose huge oval eyes looked down at them somberly.

His beard, flowing down his chest in a series of waves, glittered bitumen-black.

When they were standing close to the blaze and the young attendants had vanished into the back of the House with the barley, the priest began, "If I understand rightly, you want to build a wall."

"Yes."

"Around the Kulab *and* the E-anna?"

"Yes."

"One precinct would enclose the two? That would be a very long wall, Merkar, over canals and streams as well as fields. From here the E-anna is two *uš* away, a ten-minute walk. You speak of a wall at least eight *uš* in length, forty minutes to walk around. A long wall indeed."

"Yes."

"How could this be done? Our tenants working the Kulab land could not do it, even if we asked for help from all those who dedicate a portion of their harvest to the god."

"I will build the wall myself," Merkar said simply.

And so it was that the Wall of Merkar was built, for this was the age of heroes.

Agriculture encouraged increased population in a spiral of increasing returns.

Just as burial is memory, so space, changing visibly, becomes memory. The wheel that is time rolls across, leaving behind a line straight as a plowed furrow into the past.

Time spatialized like this, took a new form that at first asked, then insisted, and finally coerced men into searching for ways to understand it. Because time no longer repeated in the same ways, the world was unpredictable. Farmers (and more and more of the peoples of the world had taken up farming) increasingly feared and depended on unreliable external forces like rainfall and temperature, insects and storms.

One answer to the uncertainty was to amass a surplus against contingencies.

Such delayed returns demanded sacrifice today. It demanded *planning*.

But others plan, too. Those on the outside, those with no fields to plow, or who were hungry, could imagine the grain piling up in their neighbor's house. Taking it would be easier than growing it.

So Houses built walls and reinforced the windows against burglars. Temples, the Houses of the gods, would get walls.

But cities, like the wealthy Houses, would also need walls to protect its homes, so walls became known for their height and strength. Though not the only invention of the city, the wall was for a long time one of the most important, and most visible emblems of a city's reputation.

The House of Heaven

The people of the countryside look to a central city for goods unavailable locally, advice, and management of large-scale irrigation and construction projects.

Since a god owned every city, and had a house inside it, the first place to ask for help would be the great house of the god, the temple.

In Unug, also called Uruk, the home of the goddess Inanna was the E-anna, the House of Heaven.

The girl named Bilga stood, feet apart, hands on hips. Her dark eyes blazed with anger at the devastation the river had inflicted on her spring harvest.

Alal of the drooping eyelids, her tall, thin older brother, was next to her, head lowered. Ever since he lost his wife and children, he had cared little for the world around him. Now was no exception. "You're young," he muttered after prolonged silence. "And a maker of troubles, Bilga. If we still lived in Dilmun, as in the days of our father's fathers … ."

Whenever Alal mentioned the blessed isles of Dilmun, fabled paradise long abandoned and mostly forgotten, he threw his hands into the air, as if complaining directly to the gods.

Bilga would shake her head hopelessly and chide him, as she did now. "We live here, Alal, amid the watercourses of the great river. Dilmun is a dream and I will tell you why. A ten-day ago when I was sleeping under the fig tree by the house on that night of unbearable heat, I saw the gardens of Dilmun wavering, as if reflected on the surface of a canal. Inanna walked toward me through a wall of rain and stopped, no farther from me than you are. 'Nothing good comes of living in the days of ancestors,' she said. Look around you, Alal. *This* is our land. We may be the black-headed people and favored by the gods, but that does not save us from their spite." She shook her fist at the mud-streaked slope above the still-roiling river and added with barely restrained anger, "We have to do something about this ourselves."

Alal was her grandmother's sister's son, and so one of her many brothers, but he was the one with whom she lived and worked this land. Other brothers lived in other villages along the levees.

He shrugged his narrow shoulders and explained in his talking-to-a-child voice, "I tell you we would not have had to send our harvest laborers away if we still lived in Dilmun."

Bilga half-raised her fists in exasperation. The mud oozing over her sandals wheezed when she pulled her foot free. "If, if, if!" she snorted. "If you had convinced the downstream neighbors to help, the canals would not have overflowed and rotted the barley right before harvest. If you had done that, today the barley would sip what it needed instead of drowning. But no, the floods carried away the sluice at the canal head. Our brothers will harvest nothing this year, the storage bins will be empty, and even the gods will go hungry. So stop talking to me of Dilmun." She pulled a blackened seed cluster off one of the few standing stalks and threw it at him.

He turned and started trudging up the soggy slope to the house, one of seven scattered along the levees of their village. His head was down, and his steps were slow.

She followed with her eyes. He ignored the plaintive lowing of the oxen when he passed the pens. With no harvest to thresh the animals had nothing to do and were neglected. Would they still be here in the fall when it was time to plow? Would there be any grain to pay the plowman?

She sighed. His interest in life had withered. She couldn't blame him, but the darkness of his mood left her to solve the problems. "Well," she said to a stray dog sitting on its haunches a few paces away, "What shall I do now?"

The mangy black creature had been staring at her for some time, but as soon as she addressed him, he lost interest and trotted off on three legs through the date palms. A chill swept down her back. Bad omen. She followed her brother up to the house, thinking hard.

The levee above the river had drained quickly and their two young debt slaves were installing bundles of fresh reeds to replace the old doorposts. The house—indeed the whole village—still reeked of damp mud and rot.

The younger slave was slicing green shoots from the tops of the extra reeds. When she passed, he dipped his head and carried them down to the animal pens.

Alal had left the house door open. The doorway faced north, so it was cooler inside and the shade was tempting, but she knew he was already drinking barley beer and wouldn't notice if she were present, or, if he noticed, wouldn't care. "Enlil provides, Bilga," he would say, and every time he said that she replied, "Yes, this time."

She could hear him swallow and sigh. Nothing could break his gloom. She wanted to shout that it had been years, he had to wake up, to help.

She no longer believed Enlil would provide. The god had turned against them the past two years. The river had carried good soil down into the salt marshes. They still had plenty of land under cultivation, and plenty of reeds for building, but the river had overwhelmed them for the second time. She had never seen such ferocity. It tore through the sluices, undercut the embankments, and drowned most of the barley. What remained was rotting where it grew.

She thought she knew how Alal felt and almost went inside. To say what, she wondered. They had sent their relatives home. Soon, they would have to sell or release the slaves.

She was not going to let that happen. If Alal couldn't help, she would do it herself.

Fortunately the date gardens still stood, scattered among the houses. The water had not lingered long enough to harm them, and the trees to the south marched in rows along the river, towering over the orchards of fig, pomegranate, and apple. The vegetables—the chickpeas, lentils, onions, garlic, lettuces, leeks, and mustard—had received a good soaking, and thrived under their double shade. The farm and its dependents, most of the village, would not starve right away, but would surely face a slow, lingering death unless she could find a way to restore the canals. By this time next year, if they hadn't done something, they would have no choice but to sell themselves for debt.

Alal would do nothing; this she could clearly see. In his father's days the canals were new and well-made, but since then they had rotted away. Alal was still able to manage planting and harvesting but was useless with problems like this.

It was up to her and she didn't know what to do.

She walked slowly back to their house, pausing to listen at the front door for Alal's breath, its whistling intake and long, heavy sighs. It was the same every time he got into the beer. *Enlil provides* indeed!

She turned toward the river and shaded her eyes against the glittering string of floodwater lagoons in the marshy bottomlands to the east. Palm village, such as it was, fit inside a curve where the river looped lazily west before turning south again some distance away. Beyond, the steppe shimmered.

Her village was not far from the twin towns up the river, only fourteen or fifteen *uš*. There was time enough to walk to the E-anna to ask the goddess what to do. If she hurried, she could be back before dark to make Alal supper.

She passed kin who had helped and familiar folk she and Alal had hired. Closer to Unug were the houses of more recent arrivals, those who had given up their flocks or herds to stay close to town. It seemed that no matter who tended the fields, or how hard they worked, or which gods they prayed to, the crop had suffered more the past moon than any year in the memories of even the oldest among them.

They watched her pass in silence, their expressions stating as clearly as words that life was hard, and the gods asked much of men.

It would be good to stop to hear their stories, but the sun was already overhead. At the low boundary wall on the edge of town she held up her shawl for shade and hurried through the south gate into a cacophony of shouts, catcalls, braying beasts, and barking dogs.

The streets were little more than narrow footpaths hemmed by taverns, small warehouses, and the blank façades of houses. Down an alley she could see the glow of forges and hear the metal workers turning copper from the north into trinkets and small tools. Rivulets and canals small and large crisscrossed in every direction. Tributaries of the river wound in unpredictable patterns, cutting through neighborhoods and separating houses. Narrow footbridges crossed those deep and wide enough to be thick with canoes and rafts. Unug seemed more water than land.

Slaves and freemen alike staggered under reed bundles, baskets of last year's grain, jugs of oil or beer, bolts of folded cloth. Everywhere cheap mud bricks were drying in the sun, for despite recent floods, this was the dry season. Animal droppings in various stages of ripeness fumed amid the chaos.

The path climbed gently toward the city center, where the bricks dried in last year's open spaces had grown into houses. Children darted underfoot. Old women and men sat in gaping doorways swatting at flies. More and more beggars reached out their hands, if they had hands, or gave pleading, haunted looks if they did not.

Bilga missed the calm of the private vegetable gardens she remembered from her childhood. Perhaps there were some still inside the house compounds, but on the streets even the date palms were gone. Bundles of reed and other construction material leaned against every available wall, but there was no shade to be found. The city was under continuous construction.

Land should grow food, not create this confusion and noise. The gods demanded almost a third of what people grew. They may not care much for people, but this year especially, it was the most important, indeed sacred, duty to offer them something extra to convince them to change their minds and help. After all, men toiled for the gods, raised their food, housed and dressed them when asked, and entertained them when they were bored, yet the gods were seldom grateful.

"We're slaves of the gods and they do with us as they wish." How often she had heard those words!

"What's that, lady?" someone asked.

Her mind had been wandering; she had spoken aloud. It seemed she had climbed all the way to the outer gate of the E-anna!

To her left, down a road wide enough for two donkeys to pass side by side, she could see the dazzling new white of the Unug-Kulaba, home of An, the sky god, which had been built when An had founded the town and ruled over it. Ever since, it was constantly renewed, expanded, and elaborated until it loomed almost as high as the House of Inanna, at whose door she found herself. From here she could see the surrounding plain over the tangled roofs of the city. She could, she thought, see all the way to her village, even through the haze of dust in the air.

She turned. "Ah, priest, it's you. I spoke aloud?"

"You did." The priest, wearing a dark gray cloak and a threatening frown, stood beside the open gate, hands clasped inside his sleeves. He removed one and passed it over his shaved head. "You sounded unhappy with the gods, child." He was as old as he was smug.

"No, no, they're unhappy with us, it seems. But what of you? Don't you work your fields like the rest of us?"

"Like you?" He put his hands back inside his sleeves with a sniff.

"Yes, like me! I've come to see the goddess for a purpose related to the fields."

"And I am High Priest, girl. My duties consume all my hours and all my days. What do you need?"

"Ah, then," she murmured contritely. "I would like to speak with the priestess."

His lips curved into a condescending smile. "This is no festival day. The priestess would not welcome a visitor."

"Nevertheless," Bilga insisted quietly. "It is the priestess I wish to see, *Ensi*."

He stiffened. "You can talk to me. Or you can leave."

Her resolve faltered. She hadn't expected to be stopped. She always gave what she could to the goddess. The goddess had come to her in a dream. She thought all she had to do was ask. "I can't go back," she said. "We need help."

"The goddess speaks only through her *en*-priestess, and she is very busy. Come back next festival day."

Bilga did not move.

"Go on."

"Inanna told me to speak with the priestess. Please." It was only a small but necessary lie.

His frown deepened. "It is unusual to visit her."

"But not forbidden," she said humbly. "You are the High Priest of Inanna. I know I'm just a foolish young woman. I would not want to bother the High Priest with my small concerns, but I wish to speak instead with the priestess about a donation."

He started to speak, stopped, and with a stiff bow led her through a short passage into a broad open plaza filled with workshops. Stands with awnings over tables of fruits and vegetables, pots, bolts of cloth, or pitchers of dark barley beer lined the walls. The clacking of looms came from the largest building.

To the west the Kulab rose, radiant white and imposing.

She thought it might as well be a festival day, with all the buying, selling, haggling, and shouting. By the outer wall women, men, and many members of the third sex were negotiating with visitors, their hands in continuous motion as if they were telling stories. Bilga smiled, thinking of their service to the goddess. For that she would have to wait for a husband, and when was that ever going to happen with Alal around?

He led her up a steep flight of stairs to a high platform crowded with a bewildering complex of buildings. The largest, three floors above the platform, had an elaborate façade of fired brick decorated with staggered rows of black and white chevrons.

This was the E-anna itself, the é-gal or Great House of Inanna, patroness of sexuality, protector of the date storehouse, and half-owner of the city of Unug. She was also the daughter of An of the Kulab and the sister of Ereshkigal, ruler of the underworld and Lady of the Great Place, the grave to which all men return.

Her guide put out his hand to stop her at the door. "Wait here."

He disappeared inside.

From this height the courtyards below were not chaotic but lively and colorful. The distant voices were musical, the sun a little brighter and more cheerful. Even the river gleamed. It seemed the recent floods had not touched the city at all. Perhaps the gods had simply not noticed what was happening to the farms farther up and down the river.

Had the priest forgotten her? She could see nothing but gloom and the flicker of small lamps inside.

It was not the priest who finally gestured from the doorway but a slave girl much younger than herself.

Dazzled by the brilliance outside, she stopped and squinted at the Abzu, the sacred pool of sweet water just inside the door. When her eyes adjusted, she followed the girl around the basin. Passing through a narrow chamber, they came out into the calm of an inner courtyard where the only sounds were a low murmur of women's voices, the low stone scrape of their mortars, and the lilting song of a fountain in a corner. The smell of barley flour drifted over her and vanished on a vagrant breeze.

The priest was waiting there, tapping his foot impatiently. The frown under his bald head was deeper and more menacing than before. He dismissed the slave girl without a glance at Bilga.

A woman in a white tunic so long it hid her feet glided across the stone floor. Her hair was hidden beneath a large, flowing hood. Only the thick rolled brim of her official head covering was visible.

"The *en*-priestess," the *ensi* introduced drily, his voice heavy with disapproval.

The woman ignored the priest, and Bilga's understanding of the temple shifted. He was not in charge, not at all. "Bilga of Palm Town some fourteen *uš* south of the city," the priestess said in a throaty contralto.

"Yes," she answered softly. She added more firmly, "I am Bilga of Palm Village."

The priestess dismissed the old man with a wave of her hand.

The girl was struck with both awe and a sharp desire to giggle when he backed away with a bow

From the lines beside the priestess's mouth Bilga guessed she must have more than forty years. Her eyebrows were thick and a little shaggy, but her dark eyes glittered with friendly interest.

Bilga said, "Lady, I …"

"Yes? Speak up."

"I."

The priestess laughed. "All right, I'll speak first. About you."

"What do you mean?"

"You live with your brother, Alal, who is twenty-two years older than you. He is of little use on the farm since his wife and child died of fever. That was three years ago, was it not? No need to answer, it's the business of the House of Inanna to know what happens in her domain. You lost your harvest for the second year in a row. Of course, I know this as well. My name is Nin-Shar, *en*-priestess of the E-anna. That man who just left us, the High Priest, does not care about harvests, bolts of cloth, or flocks of sheep, he cares only about strict adherence to traditional ritual and complaining if the storehouses are not full. Of course it is others who must fill them. Now, come, let us go inside. It grows hot out here in the sun."

Bilga followed her into a cool, spacious room at the far end of the courtyard. Nin-Shar lit several lamps and indicated a cushioned corner. After gazing into the latticed fingers in her lap the priestess looked up with a smile. "I imagine you feel alone with your problems."

"Lady, I didn't expect … ."

"You didn't expect to be welcomed at the E-anna?"

"Yes. I mean, no."

"As you have seen, I know much about you. Word of you has reached us, of course, and you interest us. We have followed your story. So you are not alone with your problems."

"I didn't know."

"Two years ago, in the third month, the diviner here at the E-anna, a man of perfect wisdom, saw a horse trying to mount a cow. Such an unusual event foretells the decline of the land. This is how the gods communicate with men."

"No one spoke of this," Bilga said.

"The House of Inanna does not speak aloud of such things outside these walls."

"Did you speak to the goddess, ask her to avert catastrophe?"

"Inanna can do little in the face of Enlil's wrath. But the House was prepared for what happened." The priestess spread her fingers. "You come for Inanna's help. This was foreseen. Tell me, what would you ask of the goddess?"

Bilga looked down into the pattern of lines on her own palms. "We cannot survive another year like this last, Lady." She closed her fists. "We can't maintain our canals, even though we call upon distant relatives. Even with the whole village working together we have trouble keeping them repaired."

She looked away to collect her thoughts before clearing her throat. "Walking here today I saw this all along the river. The é-gal can call on as many people as needed to maintain their lands, and so it does well. I look across the river and see the goddess's barley is green while ours is black."

"Many show their devotion by their labor, which they give generously," Nin-Shar said. "It is only right, for she helps in return. We humans, we black-headed people, must do our share."

"I thought if we change the way we lay out our fields," Bilga began slowly.

"Yes?"

"Instead of making the ditches shorter, so the banks overflow when the river rises, letting the water stand in the fields and drown the barley, we should ... ."

The priestess cocked her head. "We should what? What is it you would change?"

The girl leaned forward. "Make them even longer. I thought if the furrows went from the tops of the canal banks all the way down to the marsh or steppe, then when the river floods, the water could pool at the bottom. The grain could drink without drowning."

"I'm not a farmer, Bilga, but that sounds promising."

The girl lowered her head in despair. "But impossible. We would have to build higher at the sluice gates for the water to flow. We could never hire so many men."

"Ah, I see. Why did you come here, then?"

"The E-anna and the Kulab feed those who work the gods' lands," Bilga said simply, as if that explained it all.

Nin-Shar tilted her head back and closed her eyes. After some moments in thought she said gently, "We do much more than feed, Bilga." Seeing the surprise on the girl's face the priestess rose and held out her hand. "Come. There are some things I would like to show you."

Bilga rose. "I don't understand."

The priestess laughed with delight. "You walked here today, not a feast day. You persuaded the High Priest to let you see me. That tells me much. You have an idea for farming I'm not sure I understand, but which was surely inspired by a god. You are what I expected, and now there are things you need to know."

They walked through a maze of chambers, courtyards, and hallways. The men standing at every door and corner nodded as they passed.

Nin-Shar stopped at a large wooden door of a kind Bilga had never seen. The wood alone must have come from far away, from the Cedar Mountains, perhaps. The door was closed and secured by a leather strap. A lump of dried clay impressed with a tiny image was wrapped around the knot.

Nin-Shar nodded at the nearest servant. He broke the clay seal, opened the door, and stood back. The priestess led Bilga inside.

The girl was stunned. The room was vast, larger than any she had ever seen. A series of wooden roof supports marched into a gloomy distance.

They walked between the pillars. The priestess pointed at shelves heaped with baskets of grain, stacks of fine ceramic plates and decorated drinking cups, bolts of woolen cloth, rows of copper and bronze figurines, bundles of fire-hardened clay sickles with wooden handles, cedar boxes filled with jewelry of lapis-lazuli, copper, gold and silver.

"What is this? I have never heard of anything like it. Smell this wood!"

"Cedar from the north, girl. Like all the houses of the gods, the E-anna prepares for uncertain times. We collect and protect what might be needed. In this we serve the goddess, who serves and protects the people in turn. This is one of the storehouses of Inanna. So you see, we do more than feed people."

"But there must be enough grain here to feed Unug for a year."

"Not so much as that, Bilga," Nin-Shar laughed. "But a considerable amount, yes."

"How do you ..."

The priestess held up her hand at the end of a row of shelves. She opened several bins holding small clay bricks. "Marks on the clay tell us what's in this room. This, for instance, represents the sixty *gur* of barley stored on these shelves."

"That's more than a man can carry!"

"Yes, and here are the marks for sheep. The sheep are not here, of course, but out on the steppe, assigned to this man, the shepherd." She touched a mark on the clay. "We anticipate an increase in the flocks every year, one lamb for every two ewes, for there are always deaths. Here is the same for cattle. Everything the E-anna holds is here. When we have to distribute things, we know what it is and how much, and we record who received it and when."

She had the servant tie the door shut behind them and mold clay around the knot. She pressed a cylinder hanging around her neck into it, leaving a small image of the goddess. "The seal of Inanna," Nin-Shar said. "No one can enter without breaking the seal or cutting the cord. Any time something is taken out or brought in, we make new records. That way we know how much we have and where it is."

"Surely it is Enki, owner of Eridu City who has such magic, as befits one of the Great Gods."

"Not magic, girl, but the unceasing labor of more than two sixties of people associated with the House of Heaven, the scribes of the E-anna.

Speak no ill of the goddess, Bilga; she is our owner and our patroness; we owe her everything."

"I'm sorry."

They were walking across the courtyard when Bilga suddenly stopped. The commotion of the court seemed to slow down, the sound to mute. She felt stubborn and angry again. "Why are you showing me this? I'm just a simple girl from the country. I shouldn't see this, I shouldn't know it. I don't understand."

Nin-Shar nodded slowly. "You came to offer to sell yourself to the goddess in exchange for her help, did you not?"

"Yes."

"You do this to save your brother, Alal."

"How do you know this? Yes, I want to give myself to the temple if you'll help him. He can't do it by himself."

The priestess looked pained. "It is the House of Inanna that asks for *your* help, dear Bilga, as you will see. As for your brother, don't worry about him, he is a simple farmer, and will do as we say."

"But what of my offer?"

Nin-Shar looked at her closely and snapped, "Inanna does not accept your offer, girl. If you come with me, you will understand. Then you may choose."

Two deep grooves appeared between Bilga's dark brows. She was about to protest, but the older woman had already walked out through a small door.

Bilga closed her mouth, and after a moment followed the priestess out into the midday streets.

Temples controlled land through original ownership, offerings, or acquisitions from debtors. With power came responsibility. For instance, a temple had to maintain a staff of priests and priestesses to train, manage, and feed slaves for large-scale textile production.

The High Priest or Priestess assisted by a religious hierarchy emerged as de facto ruler. Success was measured by how well he or she managed the redistribution of labor and farm produce. Temples acted as banks, consulting companies, and management headquarters, providing accounting services, a sales force, a planning commission for an ever-widening community, a board of directors, and a place for ritual entertainment. The priesthood's decisions had to be effective or the temple's influence would falter and allegiance could transfer to a different god, which put the gods and their hierarchies in competition with each other.

Over time local gods acquired attributes and specialties, but these were fluid and frequently exchanged, or merged. With no single unifying region-wide story, each city had its own patron. Attempts at synchronizing and regularizing the pantheon were never quite complete.

Like the stock market, a temple's spiritual power depended on the public's level of belief in its effectiveness. The higher the belief, the more followers. The more believers, the more donations and offerings. The more offerings, the more economic influence. It was a virtuous circle while it lasted.

Scribe

The most intractable administrative problem was memory. Human capacity is limited.

Faced with the challenges of collecting, storing, and distributing goods and services, someone decided to draw simple designs on clay to keep track of things from barley to sheep. A rounded triangle with two lines curving out of the top stood in for the whole ox. Such signs worked as visual reminders, easily combined with simple vertical or horizontal lines or deeply pressed circles or half circles for numbers.

Pressing wedges and lines in clay with a reed stylus was an administrative accounting tool. From this, writing evolved into a long-distance, external data storage system so useful it is now ubiquitous.

As soon as Nin-Shar and Bilga stepped outside, the chaotic sounds, sights, and smells of the city stopped them. Bilga shaded her eyes against the glare of the sun, staring down to where the darkness waited.

Several new bricks were drying before a new house nearby. A brick on top of the stack had the same marks on it as one in the inventory inside. "This is that what you call writing?" Bilga asked.

Nin-Shar said, "Yes."

"This mark is the same as one I saw inside."

The older woman smiled as if she were pleased. "Yes. It means scribe. This building will be dedicated to Nidaba. It will be a school for scribes."

"Nidaba? She's the harvest. She's grain," Bilga ventured. The grooves between her brows deepened. There were so many gods!

"And scribes. I have shown how barley and writing are connected, have I not?"

"Yes, Lady."

Slaves were pushing fired clay cones point first into fresh plaster over the bricks of the front wall. The wide ends, colored white, green, or black, formed a mosaic of zigzags and chevrons similar to the façade of the E-anna. The school might not be much bigger than a house, but the effect of the mosaics was impressive, and Bilga lingered. So much effort, so much time!

Nin-Shar took her arm and they moved on.

Unlike the southern district Bilga had passed through on her way to the E-anna, cattle pasturage and gardens filled broad open spaces between buildings.

They strolled down the slope toward the river along an endless blank façade. "This too belongs to Inanna," Nin-Shar explained, running her finger lovingly along the brick. "You've seen the inside of the goddess's warehouse. This is its outer wall."

The buildings stepped down the slope and met the city wall at the river's edge where the two women entered a disordered maze of bakeries, taverns, breweries, temporary storage shelters, and a counting house. In the middle of the wall, the Harbor Gate was more a collapsed section than a real gate. Crews of slaves were removing its old bricks to add to a long open portico where free men lounged, waiting for work.

Beyond the gate they looked out at the quays and the broad river as they strolled along the wide berm amid throngs of pack animals and porters. Some greeted Nin-Shar by name, others by title.

Boats crammed together to the other side of the river, their boatmen shouting to get out of the way. Stevedores, bent low under huge sacks of flour, toiled up winding paths toward the gate, forcing the women to step aside. The heat was oppressive, the air drowned in moisture from the river. Sheets of sweat flowed into their eyes. Bilga kept blinking and wiping her face with her sleeve.

Nin-Shar observed with a slight smile, "You've never been here before."

"Boats on our watercourse are so small, Lady; they bring birds and fish from the marshes and we give them dates and barley in return. I've never before seen boats so large, nor so many in one place. Where do they all come from?"

The priestess pointed to one of the larger vessels waiting near the pier. "You can identify a boat's origin by the banner hanging from the stern. The big one closest to shore with the green and white is from Isin. Unug trades with cities for lapis lazuli for our seal carvers, for wood, metal, and different kinds of wheat that don't grow well here. From farther up the Buranun and Idigna, beyond the northern mountains, we get obsidian. The dangers are many and a voyage can last for many months. Three generations ago Merkar's father Meshkiangasher set out for the Bitter Water and never returned. Yet, despite the dangers our traders go farther every year to find new sources."

"You speak of the Merkar who built the walls?"

"The same. His children's children now possess much land, and many of the buildings along this road. Sometimes I think their influence is almost as great as the House of Heaven's."

Bilga pointed at a crowd of leather workers side by side at the river's edge. They were rhythmically beating wet skins and chanting. Heaps of half-worked leather surrounded the tannery. In the shelter of the open-sided building behind them men were flaying cattle carcasses. The building was at least a hundred paces down river, but when the faint breeze suddenly shifted, a terrific stench rolled over them.

People around them started cursing angrily. The priestess held her cowl over her nose. "Now you know why the tannery is down the river," she murmured.

As suddenly as it had come the smell was gone, replaced by the pleasing odor of fermenting barley from the next building. "My brother from Date Village makes beer every fall for the family," Bilga said. "But nothing like this place."

A man with a cap of tightly curled black hair stared at her from the entrance to the brewery. When he saw her looking back his lips twitched into a slight smile.

"Ninkasi favors the brewers of Unug," Nin-Shar was saying. "Her beer is known far up the river and down beyond the marshes. She is bringer of solace and forgetfulness. You remember the lines of the poem, the one that describes when Inanna went to Eridu and drank barley beer with Enki?"

Bilga was staring at the man in the doorway. "Everyone knows that poem." He was very handsome. Still staring, she recited, as if speaking to him, "Enki and Inanna drank beer together / They drank more beer together / They drank more and more beer together."

Nin-Shar clapped her hands. "Very good. As Nidaba is the grain and the patron of scribes, so Ninkasi *is* the beer. She teaches the women how to make it, and the women worship her in return, as they should, for the brewing of it makes them priestesses of Ninkasi."

The man in the doorway touched his fingers to his lips, smiled, and went inside.

The largest of the quays was a reed platform that seemed to reach halfway across the wide river with boats lined up side by side, sterns toward the pier. From the curved bow of the largest hung a white banner with the image of the goddess, naked but for her helmet of layered cattle horns. "The sign of the E-anna, as you know," the priestess said, leading the way down the well-trodden path. From under the banner she called, "Lumagur! Come forth."

A man appeared from below. "Ah, it's you, priestess. You are welcome, but a little late; unloading's almost finished." Though he was a large man with an enormous belly and a full wavy beard, his voice was soft and high, a sweet melody.

"That is truly good to hear, Lumagur. What have you brought us in return for the issue of our fields, the labor of our weavers, and the skill of our seal carvers?"

The captain laughed heartily. "The goddess will be pleased, Nin-Shar. As usual, we bring fine pottery and such, but this year we unexpectedly acquired much cedar wood. Ah, the smell, a wonder. Already it's gone to the cargo shed." He tipped his head at the buildings along the levee, behind which the center of Unug was capped with the white House of An and the black and white mosaics of the House of Inanna. "More, I bought many hard stones from the mountains to the far north by way of Shuruppak. They are surely suitable for door sockets, as you asked. You will of course check the cargo shed yourself once we have the final tally. I'll meet you there later."

He left, picking his way through scattered boxes and bales and sacks still waiting to be unloaded.

The size of the harbor was beyond words. Bilga had thought Unug much smaller than it was, for in all her life she had gone only from the south gate to the E-anna and back to Palm Village. Unug's southern district was crowded and squalid—petty and small compared to this part with its movement, purpose, order, and amazing wealth. The quays marched along the shore and out of sight around the bend in the river to the north. There were many sixties of men clambering up planks to the ships' decks with their loads, handing them off to deck hands or dropping them straight into the holds, and then rushing back for more. Supervisors were shouting for them to hurry, other boats were waiting, the river was clogged, nothing was going to get done and it was almost time for supper! Everywhere goods were stacked, piled, or heaped in crates, barrels, jugs, and baskets. And the smells! They were so confused, so contradictory.

It was so purposeful! The idea of Alal in the midst of this bustle came to her.

"What are you thinking?" Nin-Shar asked. "You're smiling."

"I was thinking that Alal would take one look at this harbor and throw up his hands. He would say, 'This noise and confusion would not exist back in Dilmun.'"

Up on the berm the young man was back in the entrance to the brewery, hands clasped behind his back. She could feel his eyes on her.

The priestess laughed merrily. "What does he know of Dilmun, your brother?"

"What can he know?" the girl asked, turning her back on the man. "There is no such place."

"Ah, but you are wrong, Bilga. Dilmun is a real place. Let's see, yes, that ship there with the red banner is from Dilmun."

"I thought Dilmun was just a story told to children."

"Well, yes, there is the Dilmun of story. But there is also the real Dilmun, an island down in the Bitter Water."

"Oh."

"Despite our problems we are fortunate to have the river. Sending goods by donkey caravan is expensive and slow. Send a donkey loaded with grain and the drover will have fed it all to the beast before they reach their destination. Now come, let us eat something."

The priestess led the way up to a small tavern overlooking the main pier. Bilga glanced at the brewery entrance, but it was empty.

They sat at a table under an awning and watched the river. The sailors and stevedores looked like insects swarming around a hive, moving in patterns that were incomprehensible yet obviously with purpose.

A woman brought them bread and olives, tankards of beer, a plate of dates, and withdrew inside. They ate in silence. When they had finished Nin-Shar asked Bilga what she thought of the harbor.

"I didn't know." The girl shifted, glancing from the corner of her eye toward the source of her discomfort, but the road was so thick with foot traffic she could not see the doorway from here. "Unug is so much larger than I thought. So many people!"

"I wanted you to see this."

"Why?"

"Because the House of Inanna needs clever people like you."

"I'm just a girl … ."

"From the country. I know, but you learn fast, and a god, I believe Enki himself, inspires you. Unlike your brother Alal you don't give up. The world is changing, Bilga. Unug grows crowded. It grows more and more difficult to feed all the people. Inanna provides much but managing it all is … difficult. The walls Merkar built around the two sacred Houses so long ago are almost gone. True, the city builds new walls, but they soon crumble—you saw the Harbor Gate. A few years ago Isin and Shuruppak took in refugees from the east. We knew from the omens there would be trouble, but they didn't listen to us and afterwards suffered from plague. We had to turn away their merchants, so this year they've refused to help us. The harvest was bad for everyone, but further north the river did not overflow so much, and they could have sent supplies." She sighed deeply.

For the first time Bilga saw bone-deep fatigue that was almost like despair in the priestess's face. "I don't … " she began, for the moment distracted from curiosity about the man with the dark eyes.

Nin-Shar put her hand over the girl's. "There have been omens, Bilga. We have foreseen that cities will fight with others—cities that have things we want, cities that want things we have. If the Houses of the gods can no longer agree to exchange, no good will come, neither from other people, nor from the gods. I know this. Even the High Priest knows, though he won't speak of it. Towns as far as a *dana* away look to Unug when conditions are bad, when the land does not give up enough for them to eat, when the river god brings destruction, when Enlil is angry and diseases come. Already it grows more difficult for us to care for the people with what we have." Again she sighed. "There are so many people, Bilga. So many."

Lumagur suddenly appeared before them as if by magic. "We are ready to record the accounting, lady," he said in his high, soft voice.

Nin-Shar rose with a rustle of cloth and arranged her hood to hide her face. "Then let us go, shall we?"

The cargo shed was quite far from the brewery, squeezed between a fish market and a bird market. The living birds set up an awful racket, quacking and squawking. Perhaps they didn't like being confined. The fish market emitted a pungent odor that mingled with that of the birds.

It was somewhat quieter inside the cargo shed. Here the ship's contents were laid out on the dirt floor: long cedar logs still clad in their rough bark; a pile of stone ready for shaping; and many large square baskets of finely decorated pottery, very different from the crude plates and bowls the people of Unug made for daily use. There were baskets of metal ingots and several chests containing more precious stone.

Lumagur was waiting for them at a table in the back with one of the E-anna's scribes at his side. The scribe was a young, unsmiling, round-faced man. His dark, tightly curled hair, shiny with oil and bound by a clip at the nape, sprayed out over both shoulders.

The captain gestured at benches around the table. When all were seated the scribe removed a clay envelope from a box and held it loosely in both hands. He waited expectantly.

Nin-Shar nodded.

The scribe cracked open the envelope and began to read aloud, "Sent north with Lumagur son of Endug, captain of a ship of the House of

Heaven in Unug on the first day of the ninth month in the year of the first great flood, thirty *gur* of barley flour, forty *sila* ...”

The priestess interrupted. “Yes, Lukalla, we know what we sent with the captain. We were present when you prepared this tablet. What we want to know is what he brought back in exchange.”

“Yes, priestess, but it is customary to begin the accounting with the outgoing cargo.”

Lumagur grunted agreement. “It’s so you know I haven’t taken anything for myself beyond what is fair, Lady.”

“Very well.” The priestess winked at Bilga. “You may proceed.”

The scribe continued a litany of the weights, lengths, and volumes of various goods. He concluded, “Recorded by Lukalla, son of Lusilim, Scribe of the House of Heaven at Unug.”

He returned the tablet to the box and rolled a lump of clay from another container into a pillow shape that fit his palm. After a moment’s contemplation he took out a reed stylus, glanced at the cargo piled on the floor, and looked at Nin-Shar expectantly.

She spread her hands. “Right. You may begin, Lumagur.”

The captain cleared his throat. “As I said, Lady, we have brought back twenty cedar logs, each at least one *nindan* in length.” He turned to the scribe. “Are you getting this?”

Bilga watched Lukalla work his stylus across the surface. “I am,” the scribe replied, concluding a line with a flourish and looking up.

Lumagur continued calling out his cargo. When he had finished, Nin-Shar asked the overseer waiting nearby to assess the quality of the stone.

The overseer, a seasoned old man of little hair but boundless energy, lifted one stone after another, rolling them over and sometimes bending down to sniff at them. Finally he declared three of the larger ones to be of inferior quality, but the rest were more than acceptable.

“We will reject these three, then,” the priestess told Lumagur.

He smiled. “As you wish. We will find a use for them as anchors, perhaps.”

“Good,” she said. “Aside from them, your trade is good and will help the House of Heaven care for the people, and we are grateful to you, Lumagur. Lukalla, prepare an accounting of the captain’s share and see that he is rewarded.”

“At once, priestess.” He placed the fresh accounting tablet in a drying box and packed up the rest of his implements.

On the way back the two women passed the brewery one more time and Bilga’s steps slowed. “Someone must oversee places like this,” she said with a nearly successful effort at indifference.

Nin-Shar looked closely at her. "The family of Merkar owns the brewery of Ninkasi, Bilga. Why do you ask?"

The girl shrugged. "No reason."

The priestess smiled, and for the first time that day the weariness of her expression eased. "The young man's name," she lowered her voice confidentially, "is Merkar, like his grandfather, the builder of walls. He will make someone a good husband one day, and if that someone is attached to the E-anna, so much the better. Such an alliance would be beneficial to both."

Bilga lowered her face but could not hide her pleasure.

The day was nearly gone by the time they returned to the E-anna. The evening star, essence of Inanna as goddess of love, hovered to the west. "It's too late for you to return to Palm Village today, Bilga," Nin-Shar told her with a pious gesture toward the brilliant star. "I have had a room prepared for you here. I believe today you have learned a bit of what the E-anna does. Now you should stay with us for a while, unless, of course, you want to return to your village?"

"No, Lady, I have decided. I will stay."

"Good. Tomorrow you will explain to our overseers what you told me about the canals. There is much more you can do for your goddess, which you will learn in due time. Inanna welcomes you."

As the temple's power grew, extending trade networks up and down the rivers, the complexities of organizing labor to dig and maintain the irrigation ditches and canals also grew. Once committed to this intricate positive feedback loop, the influence of city centers was unstoppable, even if there were some who thought it was all going too far.

Talent migrated to the temples, which demanded more and more land under cultivation to reward it. Offering goods or labor to the gods benefitted the community.

There were signs of coming upheavals that would ultimately ripple around the world. Life for humankind was losing the last of its innocent simplicity. Urban complexes, with all their conflicts and growing inequalities, were trapping their populations in a human-made world, completely separated from all that is nonhuman. As with farming, there was no escape from this radical reconfiguration of daily life. Long distance trade had forged new levels of entanglement between people and the things they had come to depend on. Social relations demanded more and more of a person's time.

Sesame oil, raw or woven wool, inexpensive pottery, reed baskets and mats were exchanged for silver, copper, and obsidian from Anatolia, tin from Afghanistan, wood and lapis lazuli from Iran, and silver and precious stones from the Indus Valley in Pakistan. Many of the individuals and institutions that controlled this trade became wealthy, their status marked by larger houses, cool, well-kept gardens, precious jewelry, and high-quality food.

Not far from Unug a complex of tombs at Ur offer particularly spectacular examples of developing social inequality. Rulers were buried with dozens of sacrificed ladies-in-waiting, musicians, slaves, and animals, along with musical instruments, carts, and elaborate gold jewelry.

In this new world of social winners and losers, writing was the elite's prime instrument of power.

New Year

Not straight furrows, nor walls and corners, nor the endurance of linear time from past to future could entirely efface the underlying cycles of the agricultural year. Man, now a prisoner and no longer a partner of those cycles, marked important milestones (a spatial word applied to time) with festivals supported by temples or kings. Like coercion, judicious rewards were a useful tool for maintaining control.

   Festivals were an opportunity to demonstrate in physical form the essential values of the society and affirmed solidarity and group pride by returning some of the products of the group's labor to those who, willingly or unwillingly, provided it.

> Like a farmer he will plow the furrows,
> Like a loyal shepherd he will increase the flock.
> —*A Song of Inanna and Dummuzi*

Now that the meager winter rains had slowed, the end of Bilga's first year at the E-anna was near. The New Year Zagmuk Festival was fast approaching. Everyone—the people in the markets, the boatmen on the quays, the farmers in the outlying districts—was talking of little else. The young novices in their sleeping quarters whispered about it late at night. Yet neither the High Priest nor Nin-Shar said a word; the daily routine continued without interruption. Tension flowed through the temple complex like an underground river; a question lurked, unasked and unanswered. The older priestesses were silent. The younger dared not ask. They could only exchange appraising glances and quickly look away.

   Zagmuk was the occasion of the sacred marriage between Inanna and Dumuzi, the son of life, the Herdsman—the most important festival of the year! The fertility of the land, the very life of the people depended on its outcome.

Who, they asked among themselves, would be Inanna? And who, Dumuzi?

One afternoon as the end of the twelfth month approached, Nin-Shar called a gathering of the novices in the most inner courtyard. Now, each thought to herself, they would learn what they wanted to know.

So it was that the priestess was seated cross-legged before a semicircle of young women, those who had been pledged as novices the year before by the wealthier families. Bilga was the only one among them chosen by the priestess herself, though none knew of this.

Nin-Shar began with a telling of the origins of mankind. She spoke slowly and with great seriousness. "An," she began with a small wave of her hand in the direction of the Kulab. She repeated, "An, the greatest of the gods, Lord of the Sky, came down from on high to take the seat of authority at an assembly of all the gods. This was in the holy city of Nibru a long time ago, before there were people. At his side was Lord Enlil, his firstborn son, father of the black-headed people and ruler of all the lands, for it was he who had called for this assembly in response to the complaints of the other gods. He declared before An that the gods had grown weary of working for themselves. 'It is too much,' they said. 'This growing food and cooking, weaving clothing, and building dwellings!'

"So An took the seat of power, and the gods gathered round, as you are gathered here. After much discussion, much complaint, much argument, much contention, they decided to ask Enki to fashion a slave for them, a creature to perform the hard labor they no longer wanted to do. 'But where is Enki, our clever Enki of the Sweet Water?' they asked.

"Enki was asleep in the Abzu and did not hear their call! They called on his mother Nammu, the original ocean mother, to waken him, and this she did, she roused him.

"In answer to the gods' request he fashioned a being of clay pinched from the depths of the Abzu, and he brought it to life. It was not easy! At first it was a poor, feeble creature who could not walk, who refused to eat, but after a time and many tries, at last the being could stand on its own. After that, Enki gave him the attributes of mankind—he gave him disease, terror, old age, and death. 'It is their lot,' he told the other gods when they complained that the creature was unfit. 'This being is not a god, so its life must be short, lest he strive against us instead of for us. We will set his descendants to do our work, to toil in the fields, to herd the animals, to build the temples, to feed us, the gods. That is what this creature's children will be for.' Thus Enki spoke, and the others listened well and heard him."

The audience stirred; anxiety blew through them, and it was like a breeze from Ninlil, the Lady of the Wind, for Nin-Shar had said nothing of the sacred marriage, and they were eager to know. Instead they were hearing this story, often told.

"So it was," Nin-Shar continued, "that mankind was created to serve the gods. We, inhabitants of Unug, here in this courtyard, under this cooling tamarisk, beside this fountain, we are in the service of Inanna, the Queen of Heaven. We cook for her. We dress her in fine clothing of linen and wool each morning. We remove her clothes and bathe her at night. During the day we bring her sweetmeats and cakes, baskets of fruit, heaps of barley flour, so she may eat. We are her slaves in the same way that the women in the weaving house, the women who solicit men by the wall, and all those sold into servitude for debt, or barrenness, or disobedience, are slaves for us. It is the way of the gods."

She fell silent, looking down into her folded hands.

Still she had said nothing of the Zagmuk and the sacred marriage. The novices waited. Surely now she would speak.

Finally she stood up. "Come with me."

The young women began to speak, all at once.

"Follow in silence, please." Chastened, they formed a line after the priestess, who took a torch from the wall and led them through a labyrinth of increasingly narrow corridors. They turned left, they turned right. This happened many times, even seeming to double back or go in circles. At last, though, they emerged from a hallway wide enough for only one person into a room, empty but for a large, impassive man standing with folded arms before a blank wooden door. Around the room were a number of small stone lamps. The flames, absolutely still, were prodded into wild dancing when they entered.

As she had done once with Bilga a year earlier, Nin-Shar signaled for the man to break the lock. He obeyed without a word and stepped aside.

Before opening the door, the priestess said, "You must never speak of what you are about to see. Never. The penalty is severe, I can assure you."

The group shuddered and moved closer together.

The door swung open onto darkness. Nin-Shar handed the torch to the guard, picked up two lamps, and led them into the innermost chamber.

This was a surprisingly large space with white walls that reflected the lamplight. The priestess set down the lamps and stood back.

A form of dark honey gold seemed to flicker into being inside a deeply recessed niche.

The novices crowded forward to look. Bilga sensed its deep importance from the reverence with which Nin-Shar stepped back to allow the others, jostling and pushing, to study it. Bilga wanted more than anything to examine this glowing device on her own, but she held back. She glanced instead at Nin-Shar and saw clearly from the way the priestess stood, an island of calm amid the excitement of the novices, that this revelation of a great treasure was something the priestess had done many times over the years.

In the jittering openings between the girls Bilga could catch only glimpses of an object more than half the height of a man carved from a single block of stone, alabaster, she guessed. It seemed to glow with an inner light. Bands of delicate carving circled the outside. She could not make out details and was eager to approach, but there was no room. She glanced again at Nin-Shar, who tapped her temple: patience.

One of the novices asked what it was.

"A vessel. It is sacred to Inanna," the priestess replied.

"How old is it?"

"It is a prized possession of the goddess. We who are her servants can only see it once a year. It has been kept in the E-anna since the time Inanna founded the city."

"But that is forever!"

Nin-Shar, facing the object, held her hand out palm down. "There are five bands of carving," she observed. "Who carved it? Was it Lord Enki himself? What do you see, besides naked men? All novices ever see are the naked men."

The young women suppressed their impulse to laugh, for this was a solemn story, one they had heard in various forms since they were small, and they knew the priestess was showing this object to them because one among them would be chosen for the sacred marriage. It was the moment for which all had been waiting and the culmination of their first year at the E-anna.

Once again Nin-Shar touched her temple, and in that moment Bilga knew *she* would be chosen, that she had already been chosen, perhaps even from the moment she came to the temple the year before after the second terrible harvest. How did she know this? She asked herself and then answered that the goddess had told her through that gesture, that brief touching of the temple by the priestess's fingertips.

The memory of that harvest reminded her of Alal. She had not seen her brother since coming to the E-anna, but she knew he had used teams of oxen to plow long fields from the canal near Palm Village. When the river flooded, water flowed down the furrows as she had said, and nourished the dappled barley, and it had grown lush. Alal took credit for bringing in such excess of good fortune. He went about bragging that Enlil himself had inspired him to do this, though everyone knew the instructions had come down from the E-anna and that Alal was full of empty posturing. Even so, his surplus guaranteed no one questioned him to his face.

Bilga's disappointment was keen when Nin-Shar ushered the novices from the room and led them back through the same series of turns and hallways. At the central courtyard she dismissed them. Before Bilga could leave, though, Nin-Shar signaled for her to remain. The young women departed, chattering among themselves, and their voices faded away.

Bilga was left alone with the older woman. "I didn't get a chance to look at the stone of Inanna," she complained. "We left too soon."

The priestess surprised her. "No, this is true, you did not get the time you need. So, let us return."

They passed back through the maze. This time Bilga struggled to memorize each turn, noting imperfections, a chipped brick, a discolored patch. When they arrived once more at the chamber, she was confident she could find her way again if need be.

With only the two of them in the chamber, the figures carved in low relief on the sides of the vessel seemed to come to life. The object was taller than her waist. Its stem flared to a wide base. The body tapered up and flared outward half a forearm across at the mouth. The honey-colored stone did glow from the inside, Bilga was sure of it. She could see highlights of torch fire inside.

"Tell me what you see," Nin-Shar commanded.

The support on which the vessel stood put her at eye-level with the top band of carving where she could clearly see the goddess, and said so.

"How do you know this is the goddess?"

"She wears the horned helmet of her office. Behind her are the two reed bundles, doorposts or roof supports, her signs."

"What else?"

"She is tall, as befits a goddess, and looks down at the naked priest bringing her a huge basket of fruit. A man, whose servant he is, follows. Since he is as tall as she, he must be the most important man, a High Priest, perhaps, or Dumuzi himself. Below are other priests, other offerings: sheep, baskets, vessels shaped like this one."

"Very well. You already know this," the priestess murmured.

"The priests are naked. I have never seen naked priests bring offerings at the Zagmuk," Bilga said.

"This object is old, Bilga, from before time. Our priests are now clothed. Now, what can you say about the baskets?"

"They contain offerings to Inanna. I have not seen the ceremony, for I am young and not yet a priestess; I have never been inside the temple during the ceremony and have only witnessed from the plaza, but I believe they are offerings before the marriage of Inanna to Dumuzi."

There, she had said it.

After a pause, Nin-Shar breathed softly. "Good."

At that moment Bilga understood this was the test she must pass if she were to truly become Inanna.

She bent down to examine the first band of carving just above the flared base. As she did, she held up her hand as if to forestall any offer of help from the priestess.

"The waved lines at the very bottom are the Abzu, the sweet water from which all life comes, and on which it all depends."

The priestess remained silent.

"Above the Abzu the vessel is circled with plants nourished by its waters, the barley, the reeds. And above the plants, rams and ewes walk to the right. They must be offerings too, fed and nourished by the plants, which grow from the Abzu itself."

Again the priestess was silent.

"Above the plants and animals there is a space, and then around the middle of the vase a line of naked men walk to the left, winding through another empty space to the goddess and the others at the very top, in a different realm. The priests have barley, they have fruit, they have barley beer … ."

"Your understanding is great, Bilga, just as I had hoped."

"I would stay to study this object, priestess."

"You may stay as long as you wish, for I can tell you now that you are to be the sacred bride, you will enact the marriage rite, you will couple with Dumuzi. This, as you well know, will take place at the new moon, on the next to last day of the Zagmuk. If you accept this honor you will have much to learn. If you decline you will leave the sanctuary of the goddess immediately and forever. Choose."

Bilga did not hesitate. "I am honored to accept, priestess," she whispered humbly, and her reply was firm, yet not without fear. "I will not fail you."

And so it was that Bilga of Palm Village was betrothed to Dumuzi, a god himself, one who tended the animals, master of the sheepfold and cattle pen. And though every day she wondered who would be Dumuzi for the rite, who would lie with her on the wedding night, who would couple with her and so renew the land for the following year, she dared not ask. That he would be handsome went without saying, and surely that was enough.

In secret, though, she could not help but picture the dark eyes of the man she had seen lounging in the brewery doorway.

The following days were busy. She had to learn to pinch just the right amount of barley flour for the ritual offerings, and the chanted words, the gestures of hand and foot, of head and arm that embodied the voice and movements of the goddess herself. She had to know the names and attributes of each article of her ritual clothing; perfect the thirty-six footsteps, each one different from the others, to the bridal chamber; and practice the posture she was to take in the doorway when she received the groom.

She applied herself to these tasks as she did to all things, but Nin-Shar and the other priestesses were merciless. They criticized repeatedly, pointing out her failings, the clumsy gestures, and the places she stumbled over unfamiliar terms, the ancient syllables said to come down from Dilmun itself.

A six-day before the Festival, Nin-Shar called her aside after the evening meal. "You have learned well, Bilga, as I knew you would. When darkness falls come to my personal quarters and I will teach you the secret ways of the plow and the furrow."

The fear that roared in her ears was quickly followed by goddess-sent elation. The priestess was speaking of Dumuzi's plow and her furrow! Bilga knew well of such things for she had grown up a farmer, but she had no personal experience. Nin-Shar spoke on, but the roaring in her ears deafened her. When it subsided she asked the priestess to repeat.

Nin-Shar smiled. "Of course, Bilga. It was the same with me when I was told."

"You were Inanna at the sacred marriage?"

"I was. But I see something still concerns you. What is it? Speak."

"Must I then never marry, never have children?"

"Ah," the older women answered softly. "Like me, you ask, for I did dedicate myself to the service of the goddess. The answer to that question depends on many things, Bilga."

"What things?"

"Most of all, on whether the goddess summons you, or whether she releases you."

"And how will I know this?"

"Oh," Nin-Shar waved her hand dismissively. "You will know, this I can assure you. My quarters at nightfall, Bilga. Alone."

When Bilga arrived as appointed, Nin-Shar said, "Now we will speak of all you must know, including the secret words and caresses of seduction to entice your groom, which we call the ways of plow and furrow. The sacred marriage is a great honor for both man and woman, and a great pleasure as well since it brings both together as one in the goddess. You understand this?"

Bilga nodded.

"Then let us begin."

The day approached, and the Zagmuk had started with ceremonies and processions, music and barley beer. The novices, reconciled now to their secondary roles in the pageant, commented that the High Priest of the E-anna was filled with even more self-importance than usual, but they made sure not to speak thus in the hearing of their elders. And it was true that he was everywhere, rubbing his hand over his wrinkled bald pate and giving deep-voiced directions to slave and freeman alike. "No, no, the platters of date cakes go over there, not here! Really, have you no sense at all?" he would shout. Or he would complain that the weaving slave girls had made Dumuzi's cloak the wrong color, or the length too short or too long, or Inanna's hair ribbons too narrow. To the shoemakers he sniffed that the goddess's sandals needed another row of lapis lazuli beads, and to the junior official in charge of sacrificial animals he said that the selected ewe was blemished and to find another more suitable.

It went on like this until the day before the wedding itself, when the statue of the goddess was unveiled, and the priests on one side and the priestesses on the other walked up the great hall with offerings and laid them at her feet. To Bilga she seemed, in the uncertain light, to smile.

While the one chosen for the role of Inanna was supposed to be revealed only the day of the festival at the consummation of the sacred marriage, it was nearly impossible to contain the rumors, this year especially, since the chosen one was not a girl sent by her family but one plucked from obscurity. People whispered that she had gone to the E-anna on her own. So this year the bride's identity was an open secret.

Bilga, hidden by a window lattice, was observing from one of the side rooms of the temple. The great open court inside the gate to the sacred precinct was thronged with men, women, children, and animals. Though she had witnessed the sacred marriage at Zagmuk often and believed the past year she had come to know Unug well, the city suddenly seemed larger, more populous, diverse, and noisy. Even the smells that assailed her—beer, flowers, animal dung, ground grain, incense, roasting meat—were richer, more pungent and evocative.

Then she spotted Alal amid his plenty, for the harvest had been good. He was some distance away near the outer wall. Before him was a farmer she recognized from down the river, and she knew from the way her brother leaned in to the other man, reaching out with his hands, and from the angry tilt of his head, that he had just learned of her role in the sacred marriage.

When Alal abruptly turned and strode toward the outer gate followed by his servants and slaves and all his offerings, she knew also that he had become her enemy and the enemy of all who knew her.

Nin-Shar appeared beside her. "What is it?"

"Alal."

"He knows."

"Yes."

"Do not concern yourself, Bilga. The goddess favors you."

The first days of the festival flew by and the day of the marriage arrived. Bilga stood quietly in an inner chamber, surrounded by her handmaidens. In the eyes of the others she was perfection, an alabaster statue with reddish highlights the same honey color as the great vessel in the inner sanctum. The young women fluttered around her, taking their parts seriously for once as it was solemn business preparing the living goddess for her wedding night, and so they bathed her carefully, scraped her clean and purified her, anointed her with fragrant oils, with cedar and rose. They arranged the ribbons in her hair, carefully arranged the lapis jewel over her vulva, draped the new white linen dress over her shoulder, and clad her feet in the sandals crusted with lapis beads, with gold beads, with copper beads.

None mentioned the still carefully kept secret, the identity of the Herdsman, the groom, the beloved of Inanna. Whose plow, she wondered often, fleetingly, would break her furrow?

She quieted her nervous excitement, certain the goddess would help her and would provide. Had Inanna not come into her friend and teacher Nin-Shar when it was her turn to be the goddess? Had Nin-Shar not assured her the event was radiant with the divine presence; that it would bring surpassing ecstasy and a delight that would last for her lifetime?

The double hours passed in a swirl of colors, scents, movement, anticipation, regret, fear, hope: all the sensations and feelings she had ever known compressed into one morning, one forenoon, one afternoon, and then the time arrived.

Priestesses massed on the platform before the temple chanted the story of the marriage, how Dumuzi had asked her, how she had not answered him, how he had asked again, and again, until she accepted him, how she told him to come to her mother's house, her house, with his gifts.

Bilga remained hidden somewhere inside the temple. Few, indeed only Nin-Shar, could know that she was prostrate on the floor of the innermost chamber before the great vessel of alabaster asking the goddess to come down into her, to fill her.

When she stood, Nin-Shar was there. From the priestess's expression Bilga knew she was no longer the girl from Palm Village, she had become the goddess Inanna.

They walked in silence through the maze of corridors and passages, the courtyards small and large, until they were standing by the Abzu basin. On the other side the great cedar doors of the temple were closed, but through the thick wood they could hear the chanting of the priestesses grow louder and faster, leading up the moment when Dumuzi would be begging for entry at the door.

Everyone knew that after that Inanna would open the door, would see him there, and would call for her bed to be filled with perfumed straw, for the bedding to be spread, for the chamber to be prepared.

A roar outside announced Dumuzi's arrival. The chanting enumerated his gifts and those of his best men: the platters of cheese, bushels of barley, the roasted kids and lambs, the ducks and geese, the fresh carp from the river.

The chanting stopped. There was a pause followed by a strong knock on the door.

Nin-Shar squeezed Bilga's shoulder and melted away into the inner recesses, and Bilga was alone.

There was a second knock. She stepped around the Abzu tank, noting in passing the reflection of torches on its mirror-still surface, living flame, sweet water. At the threshold she waited. When he knocked for the third time she called out, "Enter, beloved, your bride waits for you."

In answer to her nod the gatekeepers slowly pulled open the great temple doors, folding them back against the outside wall. Sunlight poured in, setting fire to the golden ribbons woven into her glossy black hair.

There, swimming in light, was her beloved, Dumuzi, herdsman, shepherd, man, god, and as she had hoped in secret, he was indeed the man she had glimpsed almost a year before, the dark-eyed one, Merkar, a name known as far as the great city of Unug was known.

And so she stepped forward and flawlessly executed the bride's seductive dance of welcome and acceptance, after which she stretched out her hand and drew him inside. The gatekeepers swung the great doors closed amid the cheers and catcalls and stamping feet of the great crowd of witnesses outside.

The tumult gradually subsided and the crowd fell silent, anticipating word of a successful consummation.

Stars burst out, brilliant and unapproachable. The evening star, the star of Inanna, burned down and vanished out of sight, as Inanna had done into the temple.

The people waited long into the night, exchanging increasingly anxious glances, but at last, some time after the first watch, they learned with a roar of approval that all was well and the closing feast in honor of the holy couple would begin at midday. Already the cooks were at work, sending aromatic smoke into the early morning air.

Indeed, the gods, those non-ordinary entities with which humans had long shared the world, were behaving more and more like ordinary men and women. Complex city life gradually drew them down into the growing built world, where they, too, were increasingly estranged from the wild. They followed people into their houses and courtyards to meet, discuss, eat, drink, quarrel, make love, die, and sometimes come back to life. As it was on earth, so it was in heaven, except, of course, for the dying and coming back to life.

Having humanized the gods, people commemorated them in artifacts of great beauty like a fourth millennium object called the Uruk Vase. It was over three feet high and weighed six hundred pounds, yet thieves managed to snap it from its base during the looting of the Baghdad Museum in 2003 and carry it off. A few months later, three unknown men returned it—in several pieces.

The figure of Inanna shares the top register space with the one bringing her offerings and his servant.

These nearly human gods could only impress and dominate through position, size, and, of course, immortality. The goddess could be an inspired human, a statue, or an image. To the Sumerians she and her fellow divinities were surely real and tangible. Their favor could determine the course of human lives and the success of the harvest, always at a cost.

As with most early cultures, the Sumerians had no word for religion.

Borders

The city of Unug grew wealthy in land and cattle and amassed great riches, but with growing prosperity came new problems. Neighboring cities were also expanding.

The absolute power of the temple would not last forever. Sooner or later other cities would challenge it in open conflict, which called for a different kind of defense and the creation of a new center of power, the royal palace and its military might.

Merkar, grown stocky and gray-headed, nudged the animal with the toe of his boot. The swollen, flyblown flesh gave a little, but the donkey was decidedly dead. "Killed," he muttered.

Bilga scanned the steppe as if she could spot the killer somewhere in the emptiness. A line of mountains formed a faint smudge to the east. Otherwise she saw only scrub, a far-off herd of antelope, and a long, sleepy snake coiled loosely in the scant shade of a hummock of dry grass a few paces away.

She smiled, a bit grimly. "The arrows would have told you that."

The arrows in question, seven of them, protruded from the donkey's left side. The animal's tongue lolled, and its eyes were open. It looked surprised.

Merkar grunted. "Yes, the arrows."

"Whose. The Amurru? God of Drought? God of the Nomads?"

"No god. Real arrows. Not the work of nomads either. Nomads hunt antelope like those over there, not donkeys, and they certainly don't waste arrows."

Earlier that day a shepherd at one of Merkar's farms told them of the slaughtered donkey, and after the midday meal they had walked out to see for themselves. Now it was late, and their shadows reached out alongside that of the dead animal toward the distant smudge of mountains.

"If not nomads, who?" she asked.

Merkar swept his eyes across the empty plain. "Can't say," he murmured. "I don't recognize the arrows."

"There must be signs up and down the river. Recent crop failures that would send desperate people to hunt our livestock."

"I sent runners north and south, but it will take time."

She took his arm. "Come, let us return."

They started toward the farm. The only sign of mankind in all the circle of the world was the smoke of a cooking fire at the farmhouse rising into the hot, still air, a dark line against the dazzle of the setting sun.

"A city, then. Larsa, perhaps?" Bilga suggested tentatively, shading her eyes against the light. "It's not so far and I've heard … ."

"Larsa belongs to Utu, the Sun," he objected, and she imagined the smile in his voice, knowing she thought of Larsa because they were walking together into Utu's light. "Of course, Larsa has ties to Isin, but why would they make trouble for us?"

"Because Isin is still angry we turned their merchants away during the plague more than twenty years ago?"

They trudged on. Merkar cleared his throat and spat out the dust. A scorpion darted from underfoot and disappeared into the scrub. "Your brother," he began.

"Alal is old, Merkar. He sees how Unug prospers and thinks Palm City should prosper as well, but he's lazy and his harvests are small while the canals of Unug extend many *dana* in all directions and are the envy of the world."

Merkar put his hand over hers where it rested on his bicep and looked at her fondly. "Because of you, Bilga. You have guided our fortunes as much as I. So they depend on you as much as on me."

She dismissed this. "Then thank the goddess."

"Of course. We work hard, both of us, but your long furrows saved many a harvest for us all these years." He laughed aloud. "They are the reason Unug trades all the way to the mountains to the northwest, and to the mountains to the east. They are why distant cities there send us metal and stone in trade. They are why our family now has many fields and we can live by the sacred compound in Unug. They are why we cooperate closely with the E-anna. They are the source of our wealth and power."

"Palm City also grew," she observed. "More than twenty houses now."

"Yes, sometimes Alal's harvests have been blessed by Enlil and by Inanna," he replied. "But Palm City has no god's House, and, as you said, little surplus. I'm not afraid of your brother, Bilga. Palm City will always be a poor place beside Unug."

"Yes." Though she said nothing further, she knew how much Alal resented her husband, how hooded and cunning her brother's eyes were when he looked at him. Despite his years, Alal, she knew, was still someone to watch.

When they arrived at the farmhouse they found five strangers, burly boatmen from the southern marshes, seated in the shaded garden sipping barley beer offered by Merkar's chief overseer. They stood respectfully when Merkar and Bilga appeared, but Merkar waved his hand. "Sit, sit, I see you are already refreshed."

He clapped his hands for beer for himself and Bilga, and when they had bowls beside them, long sipping straws in hand, and dates and olives on plates nearby, they spoke together of crops (more fields of sesame than ever, the oil was much in demand in the north) and weather (hot, always too dry, not enough winter rain). They spoke of families, of who had children and who had died in the past year. Merkar and Bilga's own children, Zi and the girls, came out and were presented, and then were sent away.

When it grew dark, a slave installed torches in the garden. It was then that Bilga signaled to her husband. Merkar nodded and asked softly of the visitors, "What news, then?"

The eldest understood the change in tone. "We've been to Larsa, Merkar. We've been to Ur, to Umma and Adab. We saw the great House for Enki you're having built at Eridu. The name of Merkar is known from the Bitter Water in the south to Nibru and beyond in the north. It is said that Merkar, whom some now call Lord, *En*-Merkar, gave to the world the marks on clay that record all our transactions. They say Enmerkar is the son of Utu himself."

"They say all this, do they?" Merkar waved away such flattery. "Perhaps they speak of my grandfather, who built a wall around the sacred compound of Unug-Kulab and the E-anna. I have done little but send traders to the edges of the great world disk. So what you say is fine to hear, but there is something I must know that you have not yet spoken."

The eldest bobbed his head thoughtfully. "Your overseer told us of a donkey filled with arrows."

"Mmm, yes."

"It is said this donkey was killed by a god. By Amurru, perhaps."

"This particular donkey was killed by people. With arrows, yes, but the arrows of people I do not know."

The elder, as though he had lost control of his head, blurted, "Cities, it is said, grow jealous of Unug, of Unug's power, and what Unug has, what the E-anna has, and the Kulab of An. Great treasures fill the warehouses of the temples, it is said, food enough to feed great cities, mountains of wool. So we have heard."

And here it was. "Of which cities do you speak?" Merkar asked, more softly than before.

The elder recited a list, and it was long. Larsa was among them.

Later, when the visitors had left to sleep on their boats, Bilga urged caution. "These may be just stories carried around by boatmen. Before we act, my husband, we must know more, for if you find that what they say is true, you must request an assembly in Unug. If Unug loses its power, we lose ours. I sense that this time you must act. It will not be easy to chastise those cities. The E-anna will help, of course, but our strength is in our scribes, in our knowledge, and our contacts. I fear, Merkar, this dead donkey in our fields is the beginning of the great changes Nin-Shar foretold to me."

The runners returned with more disturbing news: Isin, Larsa, Umma, Lagash, Anshan, and Adad were assembling men. "They plan to take what is not theirs. You know what they say, Merkar: 'He who owns many things is constantly on guard.' It is an old proverb, more true now than ever."

Though there was little time to call the assembly, many prominent landowners and craftsmen gathered in the open space between the Kulab and the E-anna. Several priests and a few priestesses from both Houses stood on the temple platform, hands clasped inside the long sleeves of their gowns. Behind them the great doors to the Inanna Temple were flung open. Torches winked in the darkness of the entrance chamber.

Merkar climbed the steps, bowed his head at the line of priests, and turned to face the many sixties of men gathered below. Bilga, High Priestess of the E-anna despite her marriage, stood to one side, hands clasped, head bowed, immobile. No one could see how troubled she was.

Merkar spoke in a loud, firm voice. "You've already heard we face a threat. Many cities, Larsa, Umma, and Lagash among them, plan together to attack Unug. They would loot our storehouses. I don't doubt they would violate and carry off our women."

The crowd murmured in disbelief.

"We cannot allow this," Merkar said.

"What should we do, then?" someone shouted.

"We must prevent them from reaching the city," Merkar thundered. "We will go out to meet them."

An uneasy silence followed this statement.

The same man shouted, "What can they do, these people? There is no city like Unug. We can easily turn them back when they get here. Unug is the largest city in the world, the richest, and its people are many. Why should we abandon our fields, our animals, because of some rabble from smaller cities, cities not favored by An and Inanna?"

"Because some part of this rabble, as you say, is already killing animals belonging to the people of Unug. Most of you, those with lands to farm, have lost sheep, or goats, or cattle. Is this not true? Raiders have destroyed them not for food but out of spite or for revenge. And there is more. You, merchants of the sacred compound, and you potters and metal workers, if our city is harmed, what then will you have to eat but dust and ruin, what will you use to make your pots, your tools, and with whom will you trade?" He stepped back, allowing the assembly to discuss.

Many agreed with Merkar, saying what he said was true, some among them had lost animals and had had fields burned for no apparent reason. Others protested these were small raids, a nuisance, of course, but not really serious. It had all happened before when harvests were bad.

Soon everyone was speaking at once, and the assembly fell into disarray. Men shouted at their neighbors, calling for immediate action, while others pleaded for calm. This would pass. There was planting for next year's harvest coming up, they would have to tend their flocks! Yet others shouted no, these strangers would steal from our herds, our flocks, they would burn more of our fields. And our women, what of them!

Merkar watched the tumult in silence before moving to the side of the platform to stand beside the High Priestess. "What do you think?" he murmured.

"Let them talk. Before supper they will grow weary, and when they do, they will turn to you."

He was surprised. "You truly think this? How can you know?"

"I have been a priestess at the E-anna for more than twenty years, Merkar. In that time I have seen many assemblies."

"As have I, Bilga."

"Remember, then, what happens. The people contend, each according to his interests. That is what an assembly does. When they grow weary, they become like children and turn to a father. You are that father, Merkar, as your grandfather was the father of the city when he built the wall around the sacred precinct. The city looked to him for advice, to tell them what to do, and they will do the same with you, for you are Merkar. Or, I should say, Enmerkar."

He lifted his head, stretched his neck with a deep sigh, and waited, unmoving, as the assembly chattered and bickered.

Soon enough it came to pass as Bilga had said. A skinny farmer named Shennur, known to all as intelligent, if quarrelsome and quick-tempered, stepped forward and called. "We have discussed this matter and have decided it shall be as you propose, Merkar. Those with land or wealth in wool or metal will hire workers to go with you, to fight if they must. Some will send slaves. If the E-anna and the Kulab provide men as well as supplies, we will send out as many of such men as you require."

"Excuse me," Merkar replied. "I require none. I will contribute my share, of course. But I am not to lead."

"Someone must lead, Merkar. Someone must decide which way to go, and how far, and where to confront the invaders. We have spoken together, and we have decided you are that man. It was you who called this assembly, you who warned us of the danger. It is you who should lead."

"No, Shennur, I'm a simple farmer and trader like my father, and his fathers before him back to the time of Meshkiangasher and before. I am a simple man, and just one among many. I see several here, prominent men, who could undertake such an expedition." He pointed them out, one after the other. "You," he said, calling their names. "Or you."

They shuffled their feet uneasily, their heads lowered.

Shennur spoke again. "We ask that you take up this task, Merkar. The people of Unug have decided you are the best for this task. It is an honor you cannot refuse."

Yet Merkar again refused, and for a third time Shennur pleaded, and this time Merkar accepted, and so became the *lugal*, the great man in charge of an expedition against the enemies of Unug, as was his intention all along. "But when this matter is settled, I will put down the burden you lay on me this day," he said, and they agreed.

Being *lugal* was an honor that would never leave him. Therein lay more power than a man had seen before.

Thirty days later, Merkar was at the main counting house at the E-anna overseeing the seemingly endless lists of men, rations, pack animals, and other material in preparation for the expedition. He was telling the foremen how many baskets of barley to pack and how many boats and boatmen to hire for the duration of the campaign when Bilga appeared. She watched him pretend not to notice her. He soon gave up. "What is it?" He realized this sounded abrupt, so he rewarded her with one of his dazzling warm smiles. The smile had gaps in it now, but the remaining teeth were white as when she first saw him standing in the doorway to the brewery.

"The Diviner sent word," she replied. "It is time."

"Very well."

Merkar and Bilga walked back to the temple. On the third floor they came out onto a terrace overlooking the river where the diviner, a tall, spectrally thin man with prominent cheekbones and deep lines enclosing a thin line of mouth, stood beside a long counter covered with the implements of his art. There were bowls of flour, water, and dried bones; there were small animal organs, kidneys, livers, and spleens. At the far end was a stack of clay tablets Bilga knew contained astronomical tables. A slight breeze off the river had set the flames in a small brazier dancing. Above the horizon a few clouds drifted over the uneven rim of the mountains.

Bilga well knew there were other methods of reading the world. Had there been any recent streaks of light in the night sky? Any rumors of animals aborting spontaneously, unusual sounds, birth defects?

It was very quiet despite the breath of the air. Only the indistinct cries of boatmen far below reached them; their voices seemed to make the silence even more impressive. They looked like insects trapped in amber from the far northern forests.

"You are ready?" The diviner's voice was deep and sounded hollow, as if he spoke from a deep well. Bilga suddenly imagined he had fallen into one that reached all the way down to the Great Place where Ereshkigal ruled. No well in this part of the world was nearly so deep. The image almost made her laugh, but she kept her composure.

Merkar rumbled that he was ready, so get on with it.

The diviner looked over his counter and reached for the bowl of flour. He was about to pick it up but changed his mind and was reaching for a flask of oil when a bright green bee-eater bird landed on the balustrade overlooking the river. Its yellow chin, black eye patches, white forehead, and red wings gleamed as though waxed. It stared at them with intelligent brown eyes, head cocked to one side, a still-struggling bee in its long sharp black beak.

The diviner stopped moving. His hand dropped to his side. He and the bird locked eyes. Then, with astounding deliberation, the bird snapped its beak and swallowed the bee. After staring for some moments longer at the diviner it swooped off the balustrade with a series of loud chirps.

The diviner staggered back a step, mouth agape. The others watched the bird dive toward the river and shrink in size until it disappeared among the boats.

"What is it?" Merkar asked. "What does it mean?"

Bilga said, "Very late in the year to see such a bird."

"I know that. Well?"

The diviner straightened and for the first time a smile touched his lips. "A good omen, Enmerkar, very good. The bee is the enemy of Unug, and you are the bird, my Lord."

"I see. Yes, that is good. Anything more you can tell me?"

"Would you have me look at the flour, or the oil, perhaps? The flames are moving, and those clouds out there … ." He waved his hand vaguely at the horizon. "But omens come in many forms, and this message was very clear. The clouds appear on the right over there, and that confirms it."

"Very well." In answer to Bilga's approving smile, Merkar gestured for one of his attendants to bring the diviner his reward.

And so it was that Merkar set forth with four sixties of men carrying sickles and knives, axes and clubs, and they met the men of the rival cities, and after a great struggle in which many died on both sides, the men of Unug turned back a great evil.

Once the jubilation of the victory celebration at Unug was over and life resumed its former ways, Merkar and Bilga tried to take up their private lives again. But putting down the burden was more difficult than taking it up, and there was a sweetness for both of them in carrying it.

Because Merkar already knew how to settle disputes and regulate boundaries between fields, and to record them for days yet to come, Shennur and the others insisted he should be the one to settle with the neighboring cities, especially the ones expanding as fast as Unug.

So instead of inspecting his fields and making contracts with herdsmen, and sending traders to the rim of the world, he had to decide where the boundaries between the cities lay, negotiate agreements, and have the E-anna scribes record them so all parties would abide by them and agree to penalties for failure to do so.

These agreements then disturbed other cities in the vicinity, and he had to placate or intimidate them, sometimes by marching forth to punish.

In return the city rewarded him. It acclaimed Merkar not only *lugal* but overseer of the city. They also insisted that one of his sons must be ready to take up the burden when he died.

Sometimes agreements, he thought, might hold for many generations, and sometimes one party or the other would breach them and the talk, the palaver, the disputes, would rage on. "There will be no end to it," he said to Bilga. "It's almost easier to fight."

She replied with a shake of her head that it might be easier to fight, but it was far better for the people of Unug to have fields fat with barley, and storehouses filled with wool, and children who could grow into people without fear. "But," she added, looking at her youngest grandchildren playing in the courtyard outside the door, "that cannot always be, as was foretold to me."

"Yes, by the priestess Nin-Shar. So you told me many times, Bilga. You are my right arm and my voice, so if you believe her, I am certain she was right. Already there are those at Eridu who, even though I rebuilt the temple of Enki for them, would make alliances with other cities in the south, alliances against Unug. As for the big men at Lagash …."

"We have only today, Merkar, so let us live it. After all, 'Life,'" she quoted a proverb already ancient, "'is largely better than death.'"

He laughed, but she felt the bitter taste of it. Death was already hovering near, like the bee-eater bird on the terrace. "Of course," was all he said.

So Merkar and Bilga spent their days receiving visitors from other cities or sending emissaries. Soon they had more than a sixty of scribes dedicated to recording contracts and agreements. Their house expanded to accommodate a growing staff, to provide audience halls and gardens, courtyards and dining rooms, just to impress the growing parade of supplicants.

The house, surrounded by storehouses, workrooms, and scribal quarters, grew so large and rich in decoration that it rivaled the temples of Inanna and An.

When city-states grew large enough to come into conflict, power uncoupled from the temple and became secular.

A man named Merkar appears in several early texts. Real or imagined, he was an important figure. Enmerkar became the first king of Unug. This was long after his grandfather vanished into the sea. The word *lugal*, "great man," came to mean king once those who took temporary charge of waging violence forged permanent hereditary dynasties.

Lineage now extends backward beyond remembered lifetimes into the supernatural. Merkar supposedly ruled over four hundred years. Or perhaps nine hundred. This impossible span resembles the lives of the early Old Testament patriarchs, the first of which, Adam, also "ruled" for nine hundred years. Probably several generations of prominent men named Merkar were concatenated. In this way he would link with his descendants until fantastic legends of impossible lifetimes in an idealized deep past found their way to the desert tribes who composed the Hebrew Bible. They are now part of much of the world's collective mythology.

Descendants

For tens of thousands of years human conditions had scarcely changed. Earth gave of its plenty, and people followed familiar patterns. Stone technologies gradually improved, spear throwers increased range and accuracy and the bow and arrow was faster and more flexible. Blades became finer and more useful. Until the Sedentary Divide, though, change was gradual.

After the widespread adoption of agriculture, the pace of social change accelerated. Consolidation and transmission of property consumed ever more attention, and generations often inherited a very different world from that of their parents.

After three daughters, Bilga and Merkar had a son she named Zi, for Zi was part of Dumuzi's name, and after him two more, though both died in childhood. Zi gave his parents much grief during his growing-up, for he was quarrelsome and hot-tempered, quick to fight, and just as quick to forget. But Bilga, who spent much of her time with the child of her middle daughter, a charming girl of nine, did not despair. "Your uncle is young and impulsive, Usan, but one day he will be a great king like your grandfather."

"He reminds me more of grandfather Shennur than of grandfather Merkar," the girl answered gravely. "He doesn't stop for thinking, and often speaks stupid things."

Bilga knew she should rebuke the child for such disrespect, but she held her tongue. The girl, though surely too blunt, always spoke truth, and she had to admit it was so in this case.

So it came to be that when Merkar went to the gods, as befits a king, Zi took his father's place as Lugal of Unug. He moved into the palace with a new wife and set about celebrating his elevation to manhood with quantities of barley beer and pretty slave girls playing flute and lyre. Bilga hoped one day the people might call her son Lugalbanda, but it would not be soon. *Banda* meant "to be small or junior," so Lugalbanda would become famous as the younger king. But *banda* could also mean wild or fierce, which always colored Bilga's thoughts of her youngest surviving son.

It was no surprise to his mother that he grew quickly distracted and let everyone know. His duties annoyed him, the courtiers that quickly re-formed around him were tiresome and demanding, even *life*, which was the meaning of his name, made him restless. He threw up his hands, claiming he had neither head nor patience for administration. He was, he insisted, a man of action, though for some years he had engaged in no significant actions.

It was true, the people agreed among themselves, he was lazy, arbitrary, and often cruel in his judgments. The palace, for the word *e-gal* came to signify a royal residence as well as the temple of the god, had increased in size. Hundreds of people, from cooks and butchers to fowlers and fishermen, weavers and canal diggers, worked for it under contract or lived there as staff. Hundreds more slaves worked as bearers, boat haulers, bricklayers, and collectors of clay for the scribes.

Bilga watched the city grow under her son's uncertain hand with mixed feelings. His defects worried her, but she had to admit most of his expansions had been good for the people. Unug grew larger and wealthier and its influence now extended far up the two rivers into the northwestern mountains, across the western desert to the sea, and was every year advancing farther toward the eastern mountains. And Zi, despite his laxity and callousness, was liked well enough. Or, she thought, perhaps merely tolerated; it was difficult to gauge such things from inside the royal compound, and since she could no longer easily go out alone without being recognized, she depended on reports from her own extensive networks of informants.

One day Zi came to her. She was, as was often the case, in a small garden court in the king's personal quarters with her granddaughter Usan, now a young woman, more reserved than she had been as a child, but also more observant. The court was on the northeast side of the city closest to the river, shaded by mature fruit trees, and favored by gentle breezes much of the year.

He sprawled onto a couch and started eating dates one after another with intense concentration. Suddenly he uttered, "My wife." He had long stopped using her name and referred to her only by her title as his primary consort.

"What of her?"

"She's barren. I want sons." He poured a bowl of beer and picked it up, ignoring the copper drinking straw. After a moment he drained it.

Bilga considered while he drank. "Of course you do," she said slowly. "We all wish for you to have sons. Your father Merkar, whom the gods have taken into their company, wants the same."

"So?" He stabbed a date with a tine of copper and chewed it fiercely.

"So if, as you say, your wife is barren, she will offer you one of her slave girls to give them to you. You are the king. You can have all you could want or need."

"She won't."

"She will, I assure you, my king. I will see to it. I can send Usan to her now if you want. You would talk to your father's first wife about this, wouldn't you, Usan?"

A very slight smile touched the girl's lips for a moment. She looked down. "Of course, grandmother, if Uncle so desires it." She turned to him and asked, "Is there anyone you have in mind? I know a very pretty one in the secondary kitchen I'm sure you would find appealing. She is fertile, too. Already she has a daughter of three years."

"Pah!" he growled, pouring another bowl. "Even this is not enough to cure me. I am a man who seeks action, stifled by the court. Stifled." He lapsed into silence, sipping somberly. Finally he set aside the empty bowl. "I have a plan, though. The god has spoken to me, and no, don't ask me which god. But you may be sure I will do something great for the city."

Bilga allowed no change to her expression at this despite the thrill of alarm that ran though her. "And what thing might you do, my Zi?" She emphasized his name just a little. He did not appear to notice.

Like her grandmother, Usan also showed no reaction to her uncle's change of topic. They both knew his ways, his sudden swerves.

He shifted on his couch and adjusted his skirt. Then he stretched out his hand toward the east. "The nomads. They provoke me."

"Do they? The nomads are nothing new, my Zi."

"No," he agreed. "Nothing new. But the king is confined, suffocated, and filled with yearnings. I would do something active, something great, and important, so I will go to them, I will provoke them back. I will do more. I will tear down and scatter their tents. I will lay waste to them. I will slaughter their men. I will seize their women and bring them back to Unug."

"Mm, yes." Bilga nibbled at a date. In this way she created a bond of mutual activity with her son, though her teeth were no longer strong, and she had little interest in the taste.

After Zi left the garden, pleading some kind of urgent work, Usan asked if her grandmother still, after what had just transpired, wanted her to convince Zi's principal wife to offer a slave. "It is likely she will listen. She knows of her husband's displeasure after three girl children, and she will see this as a way of keeping her place. She knows well he may just have her killed or thrown out as a barren nuisance."

"I think it would be best for you to speak with her, though we may be sure he will have forgotten all about it by tonight."

Usan did go to the royal consort's quarters and after a long conversation filled with regrets and recriminations, anger and fear, the consort agreed to let Usan pick a suitable girl from her personal household to offer her husband, the king.

When the girl came to his quarters that evening the king barely glanced at her. "You'll do," he snapped, and sent her back. "I'll send for you some day."

But it came to pass that before he remembered to summon the girl, Zi had left Unug with a large army. It was a more coordinated rabble than his father's, but still unruly and over eager.

After several months he did return with many women, and set most to work in the weaving workshops, keeping only a few of the prettiest for himself and some of his favorite advisers. By then he had forgotten all about the slave girl his wife had offered him.

In due time Usan's parents offered her to the chief diviner of the Kulab, a sturdy young man of good family with lands of his own. He came with offerings to her house and knocked three times, as was the custom, and Bilga transferred to him a suitable amount of land and animals. And a year after the marriage Usan produced an heir for him.

Shortly thereafter she shocked everyone who knew her by returning to her mother's home and sending back the bride price.

"Why?" Bilga asked the girl. "I have no need of the lands and animals. What has happened that you do such a thing?"

"I do not wish to be owned by any man," the young woman replied. "I am not a cow or fattened sheep to be bred and cast aside."

"Has Baza done or said anything to provoke such disobedience?"

Usan was shocked. "Certainly not! He is the best of men, quiet and gentle. He assures me he does not need the land or animals either, and that rightfully they are yours. So please accept them."

"If he is so kind, what could be your reason?"

"It's what I want."

Bilga could see there was no point in arguing. The girl was so very like her when she was young, and now she had found herself! "Very well." The old woman did not try to hide her smile then, and it burst forth. "If it is the way you have chosen, so be it."

It came as no surprise for her to learn some time later that her granddaughter had taken up residence in a small house behind a tavern above the harbor.

One day she hid her head and face with a soft linen veil and went down to see for herself. Usan invited her to a small inner court with a pomegranate tree, dismissed her slave girl, and served refreshment herself.

"So." Bilga looked around and had to admit the garden was cool and well-appointed, and the house was tidy, if rather small. "What of it, then?"

"I serve Inanna," the girl told her. "I give of myself and honor the goddess as you once did."

Bilga pointed out that she had married her Dumuzi in reality after the sacred wedding, and that Usan was now married to no one.

"Nor will I be again. I have some years of beauty left to devote to the goddess, I have more than enough land and wealth to care for myself and my household, and I can be very selective of the men I lie with. It is in mutual pleasure that I serve Inanna. Besides, I donate all the earnings to her temple."

"Yet you have lost status," Bilga observed. "That status, that regard, can never be regained save through marriage."

"And what of that? My son is well cared for. Baza is a good father and a good friend—often he comes down here to pass the night with me. And, though a priest, he seems to have little of either Uncle Zi's pride or the concern for reputation of that High Priest you so often told me about."

"Do you seek children to care for you when you are old?"

Usan tilted her head as though this might be a novel concept and decided it was a silly one. "No. I lie with men in such a way as to avoid having another child. I have lands and slaves that bring me comfort. My uncle is ruler of Unug. My cousin will rule after him. I serve the goddess. What use to burden myself with more children when there are already so many? And then, too, life is uncertain and often cut short. No, I will not marry again."

Usan was correct. Life was uncertain. The eastern nomads, angered by Zi's series of sudden attacks, joined forces with some of the other cities and invaded in force. Part of Unug was burned, and many were killed. Slaves were carried away, and even some of the free women.

Zi strode through the palace. He roared his rage at the images of the gods in their shrines. He insulted scribes and overseers, slave and free. He slapped his thighs with his fists and cried in pain. He swore revenge.

His mother had grown old, and when he demanded she tell him what to do she shook her head. "You, my Zi, are king and son of the king. I am only your mother, a mere woman. You might ask your great-grandfather Merkar what he would do."

That brought him up short. "What? What do the nomads have to do with him?"

Bilga turned down the corners of her month with a shrug.

Zi's brows met over the bridge of his nose and fell into a deep groove. This, everyone knew, meant he was thinking. And then a kind of light came over his expression. "Ah," he said, suddenly thoughtful. "I see. Great-grandfather Merkar built walls around the sacred precinct. Well, that's what you told me many times, isn't it? I should build a greater wall to keep them out, the nomads and the cities that have recently made themselves into enemies. Yes, I will build a wall around all of Unug, to protect the people. We will have to ask more in taxes, but it will be worth the extra expense, will it not? Yes, I will build a great wall."

It was not Zi who would build the famous walls of Uruk, though. He fell ill in the eastern mountains and died less than a year later.

The king, at last called Lugalbanda by the people of Unug, was mourned more greatly in death than honored in life, for the people said that with his death, a burden had lifted from the city.

His son, now nearly a man, proved wiser than his father, and built the walls, and they were mighty, and wide enough on top to drive a cart around the city, and people spoke of them throughout Sumer and the lands of Akkad and beyond.

Gilgamesh, son of Lugalbanda, was said to have built the longest and highest walls in the known world. His famous story may be legend, not history, but it highlights the importance of those walls. "Walk on the wall of Uruk," the epic says, sounding like a travel brochure. "Examine its brickwork, how masterfully it is built." A wall defines a city the way it once defined a house. It is a second skin around the house, itself a skin that protects and conceals its occupants.

In *The Creation of Patriarchy*, historian Gerda Lerner suggested that women "may ... have been seen as being closer to 'nature' than to 'culture' and thus inferior ... ." They were closer to nature because of their capacity to do something men cannot, something dangerous and private. In a man-built world, the capacity to bear children, so necessary to lineage and inheritance, might appear to dwarf men's accomplishments. There is no doubt that nature, associated with women and the feminine, gradually took on a derogatory meaning: uncivilized, rude, and barbaric as opposed to civilized.

Human separation from the "wild" natural world was nearly complete. Women, along with animals, slaves, and the "products" of the earth, became property to be amassed, used, and discarded. Women's sexuality had to be controlled to verify inheritance.

A king with absolute power took on the attributes of a god and ruled from the center. This was a final upending of the traditional order of man below and gods above.

This new state is called empire.

Empire's Poet

The first known speaker for empire was named Enheduanna. She was the world's first known author, as her father Sargon the Great was its first emperor.

Co-opting religion was an attempt to consolidate his power. To that end, Sargon appointed his daughter High Priestess of the Temple of Inanna at Ur.

Inscribed on a disc in the private residential quarters of the *en*-priestess at the temple were these words: "Enheduanna, zirru-priestess, wife of the god Nanna, daughter of Sargon, king of the world, in the temple of the goddess Inanna." She died in the middle of the twenty-third century BCE.

Her father was king of the world! And in 2015 her name was given to a crater on Mercury.

> You wear
> The robes of the old, old gods ...
>
> —*Inanna and Ebih*

When the shouting began outside the temple complex, even when her attendants jumped to their feet in alarm, the woman known to the world as Enheduanna barely stirred.

"They've come!" Gizi exclaimed.

Enheduanna was not worried, though she would soon learn how terrible this afternoon would be. "Do not concern yourself," she said.

Gizi was her official hairdresser and the most senior of her attendants, but still so young! How many generations of attendant priestesses like these had she seen come and go?

To reassure the other four girls huddled around her in panic she added, "They will not dare enter here."

"Here" was the recesses of the *gipar*, the innermost of the three main buildings of Nanna the Moon's enormous temple complex and the living quarters of the High Priestess and her attendants.

Her father was a great military leader and founder of the ruling dynasty. She was sister of both his successors, and aunt of the current king, Naram-Sin. Long ago, when she was still quite young, her father had sent her down to take charge of the *E-gish-shir-gal*. Now she had the power of the greatest state on earth behind her, as well as the enormous prestige of her office. Three kings had come and gone, and her power had only grown. They were safe.

Gizi pleaded that lately there had been whispers of an attempt to depose Naram-Sin. "These past few years there have been so many rebellions, Lady," the girl said. "Someday one will succeed. What if this is the time?"

"There are always rumors, Gizi." The High Priestess spread her fine hands, palms up, to show how long she had lived here in the south since her father sent her down from Akkad. Since that day her home had been here, in Nanna's House. Nanna was the owner and patron of Ur and had lived in the heart of the holy city since the beginning. Nanna ruled the cycles of time. He was the father of Utu the Sun. All the people revered him. No one would dare turn his face away.

The black-headed people had always lived in peace with the Akkadians. Most were bilingual, even, and happy enough to live under Sargon's rule.

Then Sargon declared Akkadian the official language of his new empire, and there was an ominous and uncomfortable silence in the south. He took his armies into the field and forged an empire in the crucible of war, and so unified the land between the rivers. But at what cost! The empire may stretch as far as people lived in cities in all directions, to the western sea and the eastern mountains, but it would endure only because she, Enheduanna, had succeeded in reconciling differences. It had not been easy bringing the peoples of the south into Inanna's lap! But she had done it.

She could understand how they felt; her mother was Sumerian. It was her mother, she often told those who came to the temple, that made her one of them. She had spoken their language since infancy.

All the time growing up in her father's palace, she trained in pressing the complex wedge-shaped signs into clay. Now she could write her own holy words for others, no matter how distant, to learn. Her compositions were beloved, and not just in Ur, either. She was known to the far reaches of Akkad and Sumer.

She sang her hymns at the temple ceremonies in the cadences of the south. It was as if she had been born and raised in Ur. What matter, she would say, if the dry decrees of the emperor were in Akkadian? Nanna's House was the heart of the city, and she was the heart of the temple. She made hymns in praise of Nanna and afterward she did the same for all the great temples of the empire, for Enki at Eridu, for Nippur, Umma, and Lagash. She had collected the old songs, and rewritten, expanded, and refined them, and carried them herself to the temples. From the beginning, she took her duties seriously. She studied the old religions, the old myths. She reshaped, consolidated, reinterpreted, and carefully explained them. Her knowledge of the old ways was legendary.

She had done it for her father, and for her brothers after him. Now she was doing it for her nephew. In service to the god she had lost her birth name. Everyone called her Enheduanna, High Priestess, Adornment of Heaven. Her title was on people's lips from the marshes of the south to Akkad, the capital in the north. No one doubted her wisdom and devotion. Every year she took on the role of Ningal, wife of Nanna. She walked beside the god in the processions for their sacred marriage, and the people watching her saw a goddess made flesh.

These thoughts passed rapidly through her mind as the tumult outside drew closer. Now the cries of anger and despair from the public courtyards were louder and more distinct, and she felt the first twinge of uncertainty.

She took a breath. There was no point in rushing. After all, she had often spoken about the world of the gods and how they fixed all fates. Their decrees could not be contravened. We must, she repeated often, accept and submit. To that end, she turned to her innermost devotion, her heart's solace, to Inanna, Queen of Heaven, supreme above all others.

"Lady, please, we must go," Gizi whispered at her side.

"Calm yourself, Gizi," Enheduanna said. "Inanna is with us always."

"But these men follow Lugalane, Lady. It is said he would cast out your nephew Naram-Sin and take his place. It is said his followers would restore the old gods."

"Inanna wears the robes of the old gods, Gizi. She's one of them, the supreme one. No, they dare not enter."

Lugalane was a troublemaker, nothing more. He would not enter the *gipar*. Inanna would see to it. If he tried, she would soothe him, too. Even if he proved to be a wild bull, she would soothe him. Or she would destroy him.

For once Enheduanna was wrong. The angry words, the brusque commands, the terror, came still closer. The entrance door to the *gipar* screeched open and the sound sent a shudder down her spine. Still, she remained outwardly calm. "It does not matter, Gizi. Rest easy. Our fate is in Inanna's hands."

"His men call his name, Lady. 'For Lugalane!' they shout. They are just outside. They did not stop at the gate. They did not stop in the public court. Please, Lady," she pleaded. "We must go. We can escape to the palace. They wouldn't dare attack the palace."

Enheduanna rose, fluid and graceful despite her years, which were many, for from this temple she had seen the last years of Sargon's reign, and the eight years of her brother Rimush, and fourteen years of his brother Manishtushu, and now the beginning, she hoped, of a long reign for her nephew Naram-Sin. Kingdoms may rise and fall, but the gods abide, and Naram-Sin was beloved of Inanna.

Enheduanna asked for the close-fitting linen hat with the rolled brim of her office. Gizi placed it carefully on her hair, still black, with no trace of fading. Her years were not so many, after all. Glossy braids fell beside her cheeks. The dignified effect of white and black never failed to impress those around her. Gizi stepped back and looked at her Lady with approval.

Enheduanna faced the door, hands folded. Outwardly she was serene, but her thoughts had grown turbulent. So, the rumors were true; trouble *was* coming. It was here. Well, what happened would be as Inanna decreed.

She could not help asking herself, though, where was her nephew? Where was Naram-Sin? Shouldn't he be gathering men to protect the *gipar*, the High Priestess, the temple of Father Nanna?

He should, but she almost laughed aloud, knowing he was not gathering men, not rushing to protect the temple, but was away in Elam to the east, at the city of Susa. Before he left he had told her he planned to forge a treaty there that would make any enemy of Akkad the enemy also of Elam. Whatever happened, he would not be back before summer, and this was still early spring, the wet season.

A door nearby splintered. It would not be long now.

Inanna, who could swell with rage, who could rend and smash, split skulls, slice bellies, trample wild bulls into dust, whose rage could not be stopped, could also protect. With the goddess's help she, Enheduanna, could face anything. Had not her father faced rebellions, assassins, jealous courtiers, secret plots? She could do the same.

Gizi was beside her. Enheduanna could feel the girl's tension, the trembling weakness, and gripped her arm hard.

The door to the *gipar* flew open into a confusion: shouting sweaty men, red-faced and angry; whimpering girls clinging to each other; dancing torchlight and pools of darkness in and out of hidden corners; then rough hands and a swift journey.

It was a day, or perhaps two, before the world came back into some kind of balance.

A guard ushered them into a large, clean room with a series of small windows. He was almost apologetic. "My Lord, Lugalane, bids you welcome to his home," he said formally, standing at rigid attention. "You are to be brought food and drink."

"We would bathe," Enheduanna told him in her sternest tone. "It is the way we honor Nanna. For this we require basins of hot water, soap, soft towels. And strict privacy. It is forbidden for men to be present."

The man looked puzzled. "I, I was told nothing of this," he stumbled.

"Nonetheless we require it. You know who I am? Of course you do, you abducted us from the *gipar* of Nanna. You have subjected us to a forced march … ."

"Not I, Lady!" he exclaimed, holding out his hands in shock. He seemed so small then, so helpless and indecisive. "Oh, no, not I. I was not there. I've been here on this estate all my life, never left home. But please, I am responsible, of course I am. Lugalane will kill me should anything happen to you or your women. Be patient. I will see to it."

He withdrew, leaving two guards at the door. They did not acknowledge her presence when she spoke to them, and she realized they were deaf.

She stood musing at one of the windows. They were far from Ur. This building stood amid broad fields near a canal that must be the Nanagugal along the eastern boundary of Ur. On the other side and beyond to the east lay the province of Lagash. She remembered the governor appointed by her brother Rimush shortly before he died, an unctuous man with pleasing but unreliable manners. She could expect little help from that direction.

The canal, though wide and straight, gave off a stagnant smell and was choked with floating plants through which the boats had to push their way. Some official had been neglecting his duties, probably collecting taxes and keeping much for himself. She had seen such behavior before and would speak of this to Naram-Sin when he returned from Elam.

The basins of hot water, soap, and towels arrived and were as she had demanded. The guards withdrew.

"Let us bathe now," she said to Gizi and the others.

Later, she stood at the window through the afternoon and evening. The sky darkened to lavender and then to black.

Gizi cowered at her feet, still trembling. Enheduanna put her hand on the girl's shoulder. "Fear not," she said, and knew immediately such reassurances were hollow, and that Gizi would know it. She gave the shoulder a squeeze and turned her gaze back to the empty night.

So much raged inside her! There was fear, of course. Not for herself, though there was some of that, but for the fragile state of the empire. Naram-Sin was a strong king, stronger than either his father or his uncle, but the empire was constantly beset by such rebellions as this one. What would happen to her beloved temple if someone managed to overthrow her nephew?

There was regret as well. Could she have done more? Should she have taken Lugalane more seriously? Should she have sent Gizi and the others away to the palace before the attack?

And there was anger, molten metal searing her insides. Why did Inanna do this to her? Why was the goddess not protecting her, she who from childhood had devoted her life to Inanna?

Some years before, she, Enheduanna, glory of heaven, had composed a great hymn praising Inanna's power, her terrible anger, her compassion. Did she not tell of how the goddess subdued the insolent mountain of Ebih, that false paradise of eternal summer? Had not Enheduanna elevated her above even An, the most remote god, ruler of the sky, An, the patronizing god who called Inanna Little One and then had to bow before her magnificent fury?

Inanna was testing her.

Already words were forming in her mind, words that gave form to her devotion to the goddess who wore the robes of the old, old gods. *Precious Queen, rekindle for me your holy heart.* Yes, she had been torn from her home, her daily ritual round in service to Nanna and to Ningal, his wife. She was far from home with no hope of help. *I no longer unravel Ningal's gift of dreams.*

"Lady," Gizi murmured.

Enheduanna turned with a sigh. "I will proclaim Inanna's many aspects," she said. "Often have I told my nephew he rules only with her grace. Naram-Sin will return, Gizi. He will return, and his wrath will be great, you may be sure."

"Yes, Lady." Gizi saw the priestess was looking inward and could barely hear her words.

"That man took us from the *gipar*, Gizi. He exiles us here among the dead. Do you not see how the shadows close in? He spit on his hand and wiped it on my mouth, Gizi. He would still my words. But he will not still my words, Gizi! He. Will. Not!"

Gizi managed a smile. Her mistress was strong; she was firm. She was Inanna in flesh, the storm-bull, the wind-dragon, the slayer of men! Her lady would proclaim the virtues, the qualities, the powers of the goddess so all could hear. Then the young priestesses attending Enheduanna would be restored with her to their rightful place. Their home.

"A sandstorm is coming, Gizi. Can you not taste its stony textures on the air? Can you not smell its rage? The storm-bull will trample over us."

"Yes, Lady." Now was not the time to plead for solace, to show her fear, to voice her doubts. So she concealed them all and turned her face to the High Priestess.

But Enheduanna understood. "I know you fear, Gizi. You are all afraid. But the storm-bull does not come for us." Her smile was warm and kind and filled with light. "No, it does not come for us."

On the second day, as the High Priestess had foretold, clouds of sand rose up out of the desert and darkened the day to night. Enlil of the spring winds whirled around the building, He pulled at the corners and shook the foundations. Sand hissed against the walls in a rising and falling wail.

Darkness continued for three days and as suddenly as it had arisen it was gone and there was light again. But the land outside was greatly changed. They could see people digging in the distance, looking for their houses and loved ones.

A guard brought water, bread, and meat and drink, but he trembled constantly, spilling the beer. When Enheduanna asked him what Lugalane would do with them, he backed from the room, waving his hands.

More days passed until finally *that* man, as she thought of him, Lugalane, swaggered through the door. A ragged scar on his shoulder glowed red, his mouth was turned down in a sneer, and he kept his hand on the hilt of a long dagger at his hip. He puffed out his chest, for he was short and stocky, and certain he could intimidate only by being cruel.

A servant placed a stool in the center of the room. He sat facing the women. "Well?"

Enheduanna looked at him appraisingly without answering.

He shifted uneasily. "You're going to Uruk," he said. "Get ready." He stood abruptly, knocking over his seat, and strode from the room. The servant swept up the stool and followed.

"Uruk?" Gizi repeated.

"He would hide us there where he thinks no one will find us. But this is good to learn, Gizi. We are going to Uruk, to the House of Heaven, and that is where we need to be."

So it was.

In the five hundred years since Bilga had been priestess there, the great E-anna at Uruk had grown, and fallen, and grown again. The line of Enmerkar, Lugalbanda, and Gilgamesh had faded into legend.

Though no longer the center of Sumer, Uruk was still an important city with impressive walls and a busy harbor. It was Inanna's home, and Enheduanna was her servant. Short of being home in the *gipar* at Ur there was no better place to wait for events to unfold as the gods decreed.

The *ensi* governor of Uruk was a thin man with sharp features, and very pale for a southerner. His welcome to Enheduanna and the others was too effusive, too flattering. His hands were in constant motion, pointing this way and that as he spoke of the wonders of his city. His short, unconnected spurts of imperfect Sumerian nearly made her laugh.

His father had been one of her father's generals, so he should have been loyal to Sargon's grandson Naram-Sin, but she understood that he had followed the prevailing winds and swung his support to Lugalane.

This, she still believed, was not going to be good for him. She would let him know this when the time came. And it would come, she was certain of that. Lugalane had made a fatal mistake.

The priestesses of Nanna were given comfortable, if simple quarters in his palace, built over the ruins of the palace of Enmerkar and Zi. They were confined to their rooms, given all they could want, and left alone.

The season turned. Utu the Sun pulled all sweet water from the earth, cracked the surface, and lowered the water in the canals. Animals, the ewes and goats, the cows, began to drop their young. The days grew hot and long, and no word came from the great world outside.

Enheduanna called for clay and worked on a poem, pressing the stylus and rubbing out the mistakes and pressing again, and slowly the lines grew across the tablet, following one another. *Why do you slay me?* She asked the goddess, a cry from her heart, an expression of her devotion despite her suffering.

One day Lugalane appeared again. His hair and beard were unkempt, and he seemed smaller than before. He stood near the door with his arms folded and glared, as if daring the women to speak.

They looked at him without fear this time. They had been here four months, plenty of time for Enheduanna to teach them not to show fear even if they felt it.

The strange impression that he had become smaller helped, too. So they watched him with curiosity.

This seemed to enrage him. With knitted brows he glowered at Enheduanna. "Naram-Sin proclaimed himself a god," he said at last. The tone was strange coming from such an arrogant face, almost an accusation, yet somehow pleading too. "He's gone too far. The people won't stand for it."

"What would you have me do?" Enheduanna asked, more calmly than she felt at this news. If true, her nephew had indeed gone too far; yet she understood the wisdom of it. Some people would think more than once about opposing him.

"Do?" He showed the whites of his eyes. "You will *do* nothing. You will stay here in Uruk until Naram-Sin is gone from the land. You will remain silent. I know you have been composing something, if that is what you call it. You may be sure, it will never leave this room." He turned on his heel and was gone.

The women looked at one another. "What did that mean, Lady?" Gizi asked.

"It means he's afraid. Naram-Sin returns from the east."

"How is it you know this, Lady? He said nothing of Naram-Sin's return."

Enheduanna shook her head. "You are wrong, Gizi. He said everything."

"I don't understand. We are prisoners here, but we would have heard something. The guards are always talking in the corridors. Surely what he knows is only rumor."

"Rumor flies in both directions, Gizi. He believes Naram-Sin marches against him under the sign of a god, and he's afraid. It was on his face as clearly as if written there. His only hope is to keep us here to trade for his life, but it will do him no good. My nephew will crush him, and we will return to the *gipar* as before."

Gizi concealed her doubts, and over the next days the palace in Uruk was astir with preparations. They could hear the workmen on the walls, the furtive, anxious exchanges among the palace staff.

One morning there was absolute silence. They tried the door and it swung open. The halls were deserted, so they left the confines of the palace to discover that it was as Enheduanna had foreseen. Naram-Sin had marched with his army toward Uruk. Lugalane had gone out to surrender and was taken away in a neck-stock.

Enheduanna was the first High Priestess in a tradition that lasted five hundred years. Her poem of sorrow and exile is called *The Exaltation of Inanna*. Ethnically she was Akkadian, but she wrote in Sumerian. Like

Chinese in early Japan and Latin in Medieval Europe, Sumerian remained the literary and liturgical language long past the time it ceased being spoken.

Her nephew Naram-Sin ruled for thirty-six years, his son for twenty-four, and then Ur, like all empires, fell. It had lasted nearly a century and a half. The city rose to empire again, only to collapse in less than a century.

Like all cities, the Sumerian city was a living organism, larger, more enduring and powerful than an individual. What began as the home of a god, the non-ordinary entity that gave the city its life, in time became itself the god: city and god fused and made one. It is no wonder then that when people lost their city, they, too, felt lost, abandoned.

It was an age of rebellion, of anger, of ambition achieved and hope thwarted, of great wealth and abysmal poverty. Wealth conferred privilege and comfort, but the price, particularly for women, was often confinement and isolation, while the price of poverty was servitude or slavery. The gods could ease the pain of life and facilitate the control of the elites. But these early city-states were fragile, their coalitions temporary and evanescent, their power limited. Gradually they grew more robust and more merciless. Their reach expanded—Assyria, Persia, Rome, a long sequence of Chinese rulers, kingships and democracies—but all have given way before change.

These are the risks of empire. We are fortunate we have Enheduanna to offer other insights.

But there is nothing more distressing to a society than a city's destruction.

Lamentation

The world grew rich with the corpses of cities. Sometimes the destruction was natural—an earthquake, volcanic eruption, hurricane.

More often, the destruction was man-made. The Third Empire of Sargon arose, and then it fell.

The grief is much the same.

> In order to overturn time, in order to thwart the design of fate,
> The storm winds gather …
> —*Lamentation over the Destruction of Sumer and Ur*

Rihat stood at the low railing of the sprawling barge with his arm around Ninsunu's shoulders, hugging her close. The riverbanks, broken by clumps of reeds and stretches of date palm, drifted past. The man at the bow tested and called out the depth every few meters. The water was dangerously low and progress was halting.

They were going south, back to Ur, back to their home.

For once the lands of the north had been unvaryingly calm, which only heightened their anxiety when they began hearing stories of conflict in the south. The Gutians had rebelled again, had come down from the Eastern Mountains and attacked cities and burned villages. The king, Ibbi-Sin, had repelled them, or he had failed, no one knew for sure. It took so much time for messengers to travel up the rivers, and when they did finally arrive their stories did not agree. Those in the lands of Akkad did not think it was important; the Gutians never came north.

For the entire journey down from Mari, they had seen oxen pulling plows, furrows turning, sheep bleating in their pens. Farmers looked up as the barge passed and waved.

They waved back. Sometimes they shouted wishes for a break in the drought: "May Enlil and Enki send us rains!" And the farmers would shout back, "May they send us rains."

But beneath the calm, worry nagged at them. For years now the priests and omen readers, the praise singers and masters of sacrifice had begged and cajoled, but the gods denied them. Everyone knew that Naram-Sin had once dreamed of destruction and had fallen into depression, but that was many sixties of years ago, during the first Akkad dynasty. Sargon and Naram-Sin were long gone.

Now rumors of war were so constant people had wearied of reacting. Men, they said to one another in hushed tones, were always fighting; armies were always marching. For most of the years of Ibbi-Sin's reign at Ur the Elamites and the wild tribes of the mountains threatened the empire. The king went out to fight them or built walls, but they kept coming, sensing weakness.

One day at mid-afternoon the river turned east for a short stretch, then south once more, and that was when Ninsunu said, "Do you smell it?"

They pressed to the rail, straining to see around the next bend, and the next. There was nothing to see around the first, and the second, but the stink of burning grew stronger with each heartbeat.

They had prayed to Father Nanna for a quiet old age surrounded by their children and grandchildren. This was supposed to have been their last trip to the north. The barley harvests had been bad for some years and now their barge was heavy with northern wheat for the hungry at Ur. Ibbi-Sin still had much treasure, and they were certain to do well.

They were still a day's travel from Ur when a solitary column of black smoke appeared above the distant bank, then another and another until columns seemed to leap into the sky like date palms, lines of them down the river and along the canals, etched into the thick haze.

It wasn't long before they were passing between the dying embers of farms and storehouses on both sides of the river, then the smoking ruins of a village. Dead animals lay thick in the farmyards.

They drifted through devastation and an occasional untouched farm or village, until the sun disappeared into the western desert. Then there was only darkness and that dreadful smell.

That night passed in a thick, brooding silence no one wanted to break.

They hadn't gone far the next morning before the smell of rotting flesh surmounted the smoke. Once the silence broke into distant screams: wails of pain, grief, or rage. The cries sounded so alike it was a relief to leave them behind.

Ninsunu clutched Rihat's hand. "So much smoke," she murmured. He squeezed her hand back. It was true the smoke was thicker; they could no longer see the surface of the river.

"We have to stop," the boatman said, steering to the western bank.

"We can't leave the cargo," Rihat told him.

The river-man shrugged. "Nothing I can do. Can't see the river, can't measure the depth, can't go on. We can wait for the smoke to clear, if that is what you wish."

The next morning the fires had burned out and visibility improved enough to continue, but the awful silence had returned, and the smells of decay were worse. They drifted passed Uruk, apparently untouched, and then Larsa, also untouched. Just before they reached Ur a cloud of smoke heavy with grease and death rolled over them. That was when they truly knew, though they refused to speak of it at first.

The boatman stopped in the middle of the river until the air cleared a little.

The harbor came into sight. It was deserted. A half-burned ship tilted against the main pier, making it impossible to tie up. They pushed against the derelict and fastened to it as best they could.

Ninsunu turned away with a long wail and buried her face against Rihat's shoulder. He mindlessly stroked her thin white hair, over and over, gaping at what he saw.

The city walls were breached in several places. That was bad, of course, but worse was the temple of Nanna. The temple mountain at Ur was famous throughout the empire. Even from this distance they could see that the western wall of the temple itself was gone and the others partially dismantled. Floors and interior walls had collapsed. Wisps of black smoke twisted up into an empty blue sky. Someone was burning the bodies.

Ninsunu tore herself from her husband's arms and leaped into the river. The water reached her waist. She dragged herself through it onto the bank and fell to her hands and knees, breathing hard.

Rihat glanced bleakly at his cargo and followed her. Together, wet and shivering despite the warmth of the day, they climbed the brick staircase to the Harbor Gate.

A few people had remained to pick through the bodies that littered the streets. They looked at Ninsunu and Rihat with vacant eyes. Most of the dead had been speared or slashed, and their blood, dried and caked, darkened the dirt. Snarling dogs tore at the bodies.

They wandered through the uprooted gardens, the polluted ponds, the despoiled cult buildings, the burnt trees and toppled statues, ash heaped in the doorways, canals broken, drained, and brackish, fish gasping in the remaining pools of shallow water. Everywhere flies were crawling in iridescent sheets over the dead, swirling up in dark clouds whenever the dogs disturbed them.

"Rihat!" A man hobbled toward them with the aid of a stick, one bloody foot wrapped in cloth. "You're back." His voice quavered and his bare wrinkled head was crisscrossed with cuts and scrapes. "You didn't see my son Amar at the harbor? He was on his boat getting ready to sail before the attack six days ago. I haven't seen him since."

The corners of Rihat's mouth turned down. "I regret, Laqib. There's no one in the harbor, no one at all. The quay is blocked, and we had to wade ashore. So much death! What happened? There was talk in the north, but it was hard to believe until we saw the fires, but even then we thought Ur was still strong. We were sure Ur would be standing. But now ... ."

"Nanna abandoned the city many days ago," the old man said. "Those demons from Elam broke through the walls as if they were not there. They tore in a fury at the temple. It was terrible to see. They smashed the statue of Nanna. Inanna, too, abandoned us. It is as the gods decided."

"What of Ibbi-Sin? What of the king?"

"Taken away, hands bound, barefoot, head bowed in shame. They overturned his throne and emptied the treasury." Laqib spat on the ground. "Filthy people, the Elams. They are filth and they live in filth."

"What do you see around us, Laqib? Is this not filth?"

"Yes." The old man spat again. "Elams did this, and what can we do? The gods are gone, the army is gone, the king is gone. We are helpless."

They left him muttering to himself and made their way through the ruins to their home, the compound Rihat had built when they were young and confident of the future.

They found only broken walls and burnt, fractured bricks. The floor of the room where they wrote and stored their records was littered with broken clay tablets. All they possessed, all they knew, all they were, was gone. The people of Ur had scattered; there were no mouths to eat the grain on the barge. There was no king to purchase it.

Here, in this ruined house, their children had been born. They ran through these wasted gardens, these shattered rooms. Ninsunu touched her personal amulet, thanking Nanna in a kind of grateful despair that their daughters were married to merchants in far-off cities and their two sons had taken a ship across the southern ocean and were not expected back for another year at least.

That their children were gone was the extent of Ninsunu and Rihat's joy. Nothing else remained for them but to sit down, and rub ashes on their gray heads, and try to weep.

A city fell to an invader, a plague, or a flood. The tragedy remained in memory, and some time later a scribe produced a long poem called a lamentation that described the pain of its loss, how the inhabitants came to accept the calamity, and their joy at its restoration. When a city god, in the form of a cult statue, took an important ritual journey to Nibru for renewal, her city would cease to exist for the duration. The god's return was a reason for great rejoicing.

How much worse when the god did not return.

Over half the earth's still-expanding population lives in cities. Today's catastrophes—climate, terrorism, drought, flood—seem as horrific as any in Sumer. City is home. To lose it is to lose all.

The Hebrew Tanach continues the lamentation genre. And in the famous 137th Psalm, the people of Judea lament their exile from Jerusalem: "By the rivers of Babylon, there we sat down, yea, we wept, when we remembered Zion." Then, in the final couplet, pent-up rage bursts forth: the "daughter of Babylon" will receive as she has given, and *Happy shall he be, that taketh and dasheth thy little ones against the stones."*

In the west, Jerusalem remains a potent figure of imagination. William Blake would not rest "Till we have built Jerusalem, / In Englands green & pleasant Land." The ultimate home is the City on the Hill; the City of God so sacred to three warring religions even today.

Revenge and retribution are still with us.

Usurpers throughout recorded history have named themselves heirs of their former enemies. In this way they could claim legitimacy and assuage the terrible losses they had inflicted on their new subjects. The lamentation was a way of commemorating the politics of grief, disguising the deception.

Humans are resilient, but patterns of greed and vendetta are stubbornly ingrained. We soon forget the evil we do each other, and as soon repeat it.

Afterword

Our predecessors the hominins migrated out of the African homeland and scattered around the world several times in our evolutionary history. When *Homo sapiens* followed, they found and mated with cousin species already there, including Neanderthal and Denisovan. Though we are the sole survivors, we carry them in our genome.

Why us?

An origin story we continue to tell is this: the Creator of the universe bequeathed this planet to humanity, or, rather a specific subset of humanity, and issued a mandate to seize dominion over it.

A newer version, for which we may thank Copernicus (or Aristarchus of Samos, to be picky), implies that in a sun-centric world our somewhat vaguely defined destiny directed us to the right place at the right time. Destiny is vague because we lost our place at the apex of Creation, but we were still clever and deserved all we could take.

A post-Darwinian story suggests that evolution favored us, at least temporarily. The Neanderthals had larger brains, had been around for several hundred thousand years longer, and knew the place well, but conditions changed against them. Plains and tundra opened up for large plant-eaters to wander, and our broader, more complex social organization, facilitated by language, allowed us to hunt them more successfully. We thrived in places where our cousins could no longer compete. We were, in sum, the fittest.

Here blind evolution sits in for gods. The story is a little less human-centric, a little less triumphal, but it bears the same sense of mission. Evolution simply swept humankind to prominence.

In the more recently revised version, we are a chance product of innumerable random physical and chemical encounters over billions of years in a turbulent and indifferent universe. There's nothing special about us. We may be fit today, but like many other species in a changing world, we can lose our fitness. Destiny is deceptive hindsight.

We see more clearly that a confluence of evolved traits encouraged but did not guarantee our survival. Long before the arrival of modern humans, Hominins stumbled on cooking meat. Cooking made it easier to chew and digest and more enjoyable. Because the protein was in effect pre-digested, cooking allowed our gut to shorten. A shorter gut meant less energy devoted to extracting nourishment from raw foods, freeing more energy to fuel a larger brain. Since we lacked claws and teeth and armor for an endless and expensive arms race, we had to find more innovative ways to compensate, like social patterns of hunting and exchange. We created visual art, music, or body ornaments, like Nyla's necklace, that signaled our affiliations. Sometimes trade reduced conflict. At other times, it increased it.

For some humans the long millennia foraging and wandering were challenging, but who doesn't like a challenge? For others in more favorable environments, the world offered ease and plenty. Scavenging and hunting supplemented a broad variety of plant resources. Local cultures proliferated endless variations on the basic pattern of small bands following the food. To the best of our knowledge life in the Paleolithic was relatively peaceful. It was well worth being alive despite the threat of predators, natural disasters, accidents, or occasional interpersonal violence.

In other words, we were lucky. Roughly 3.5 billion years ago matter underwent a phase shift, igniting into life. Then, somewhat later, we came along.

The most important event in our history had nothing to do with us. The Holocene arrived. It would have come even if we had not existed. Climate turned benign. Plant and animal foods proliferated. Eventually, with enough individuals and enough socially constructed interactions like initiation ceremonies, feasts, and long-distance trade, another profound shift happened called farming.

The Neolithic was the change that climate wrought. For ten millennia the climate has been a drunkard walking in an unpredictable and random path, but always trending in the direction of the farm. Nearly everyone in the world was tempted, succumbed, and exulted in the harvest.

We were, perhaps deliberately, ignorant of the vicious circle of increasing returns agriculture generated. Close, sustained communal living encouraged diseases not seen before, which shortened lives. Farming demanded more people to meet ever-increasing needs and counteract the high infant mortality it caused. Once truly begun, this cycle is difficult, if not impossible to stop.

The house became a permanent structure, usually square. Since, at this point, a house could last for generations, new concerns arose. Was the woman of the house faithful? Were her children really his? Which male son was worthy to inherit the house, the land and its bounty? Such questions provoke stress and anxiety.

Events inside the house were cut off from neighbors' eyes. Privacy encouraged doubt, envy, and gossip. Luckier or more productive farmers stored more food and seed for the following year, and wealth became a value associated with social status. Property now demanded protection.

The world was ripe for the next great innovation.

From the most productive farmlands in Mesopotamia cities evolved more complex social organizations to manage large projects. Writing, the technology for storing and retrieving information emerged. It accelerated the same positive feedback loop, favoring high status individuals with greater social control over an increasingly stratified urban society, which in turn required more and more record keeping.

Along the way, without noticing, people lost connection with the wild environment. The human-built world was carved up into straight lines, the default position for anything new. Straight lines were convenient, easy to make, and the shortest distance.

Even writing progressed in straight lines. Roads connected locations. Centers of power grew upward, reaching for the heavens. Plows made furrows that encouraged water to flow along them for irrigation. The human-made world was dominated by the straight edge.

This fact has reorganized the hierarchy of our senses. Sounds, smells, textures, even tastes did not disappear, they simply ceded their importance to vision, which now dominates our sensorium.

We traded in individual, local gods for lineage ancestors or clan totems. These gave way to gods whose residence in town managed important external and unpredictable facts of life like rainfall, plagues, underground aquifers, grain and date palms, sexuality and reproduction.

Towns merged together, often by force, into empires, and with empires came the religions of the book, first as hymns and psalms in praise of city gods, then collected tales of a people that bound them into an identity. The gods were appropriated for political purposes and did not object.

Writing has given us history. New knowledge leads to new insights. Interpretations change. The lesson for today is that agriculture, like all broad social changes involving physical technologies, brought a constellation of unintended consequences. We used our power to change the Earth for our benefit, to terraform the planet, and shaped our views to support our actions. Today man-made (anthropogenic) change threatens us with the possibility of self-extinction.

Just because we could do something does not mean we should. As Sir David Attenborough suggested, "Maybe it is time that instead of controlling the environment for the benefit of the population, we should control the population to ensure the survival of the environment."

Agriculture has bestowed gifts, certainly. Why would we have accepted it if it hadn't? But there is a cost and the bill is coming due. Prophets have repeatedly warned that time is running out, but according to conventional wisdom the warning has always been wrong, so why would it be right this time?

The problem with conventional wisdom is that although it's nearly always right, sometimes it's not. One day, as ecosystems begin to fold, the prophets could get their due, but, as an oft-paraphrased Danish proverb suggests, "It is difficult to predict, especially the future."

There are positive signs. Populations in industrialized nations are beginning to decline. Technologies to alleviate climate change are expanding faster than projected, though caution is advised: don't forget those unintended consequences. The drama isn't over until it's over.

We cannot revert to hunting and gathering; there's too little wild left and we've lost too much knowledge and understanding. Our intelligence got us into this mess, and instead of working together, we're increasingly at odds.

It's past time to evolve a global culture that fosters wellbeing and curtails greed. We will need dedication and good will, but if we can reach a broad agreement on solutions to the hard problems we inadvertently created, and we can be prudent in our application of them, we may yet survive, even prosper.

Over millennia, organized masses of modern humans changed the world and invented gods to guide them. Some of those gods have not served us well. We built empires and multitudes suffered. We were shortsighted. Predicting is hard, but now is the time to face the future with open eyes, aware always of the lessons of the past. Though we will certainly make new mistakes, we should, at least, avoid old ones.

Suggested Reading

Clottes, Jean, and David Lewis-Williams. *Shamans of Prehistory: Trance and Magic in the Painted Caves*. New York: Harry N. Abrams, 1998.

Crawford, Harriet. *Sumer and the Sumerians*, 2nd ed. Cambridge, UK, and New York: Cambridge University Press, 2004.

Finlayson, Clive. *The Humans Who Went Extinct: Why Neanderthals Died Out and We Survived*. Oxford: Oxford University Press, 2010.

Flannery, Kent V., and Joyce Marcus. *The Creation of Inequality: How Our Prehistoric Ancestors Set the Stage for Monarchy, Slavery, and Empire*. Cambridge, MA: Harvard University Press, 2012.

Guthrie, R. Dale. *The Nature of Paleolithic Art*. Chicago: University of Chicago Press, 2006.

Guthrie, Stewart Elliott. *Faces in the Clouds: A New Theory of Religion*. New York: Oxford University Press, 1993.

Hayden, Brian. *Shamans, Sorcerers, and Saints: A Prehistory of Religion*. Washington, DC: Smithsonian Books, 2003.

Hodder, Ian. *The Leopard's Tale: Revealing the Mysteries of Çatalhöyük*. New York: Thames & Hudson, 2006.

Jacobsen, Thorkild. *The Treasures of Darkness: A History of Mesopotamian Religion*. New Haven: Yale University Press, 1978.

Kripal, Jeffrey J. *The Serpent's Gift: Gnostic Reflections on the Study of Religion*. Chicago: University of Chicago Press, 2006.

Leick, Gwendolyn. *Mesopotamia*. London: Penguin, 2003.

Lerner, Gerda. *The Creation of Patriarchy*. New York: Oxford University Press, 1986.

Lewis-Williams, David. *Inside the Neolithic Mind: Consciousness, Cosmos, and the Realm of the Gods*. London and New York: Thames & Hudson, 2005.

Mithen, Steven. *After the Ice: A Global Human History 20,000–5000 BC*. Cambridge, MA: Harvard University Press, 2004.

Weisman, Alan. *The World Without Us*. New York: Thomas Dunne Books, 2007.

Whitehouse, Harvey. *Arguments and Icons: Divergent Modes of Religiosity*. Oxford and New York: Oxford University Press, 2000.

Witzel, Michael. *The Origins of the World's Mythologies*. Oxford: Oxford University Press, 2013.

Wrangham, Richard. *Catching Fire: How Cooking Made Us Human*. New York: Basic Books, 2009.

CPSIA information can be obtained
at www.ICGtesting.com
Printed in the USA
JSHW042331200121
11107JS00005B/68

9 781789 206203